Metabolic Syndrome and Psychiatric Illness

Metabolic Syndrome and Psychiatric Illness

Interactions, Pathophysiology, Assessment and Treatment

Scott D. Mendelson, M.D., Ph.D.

AMSTERDAM • BOSTON • HEIDELBERG • LONDON • NEW YORK • OXFORD
PARIS • SAN DIEGO • SAN FRANCISCO • SINGAPORE • SYDNEY • TOKYO

Academic Press is an imprint of Elsevier

Academic Press is an imprint of Elsevier
30 Corporate Drive, Suite 400, Burlington, MA 01803, USA
525 B Street, Suite 1900, San Diego, California 92101-4495, USA
84 Theobald's Road, London WC1X 8RR, UK

First edition 2008

Notice
No responsibility is assumed by the publisher for any injury and/or damage to persons
or property as a matter of products liability, negligence or otherwise, or from any use
or operation of any methods, products, instructions or ideas contained in the material
herein. Because of rapid advances in the medical sciences, in particular, independent
verification of diagnoses and drug dosages should be made

Library of Congress Cataloging-in-Publication Data
A catalog record for this book is available from the Library of Congress

British Library Cataloguing-in-Publication Data
A catalogue record for this book is available from the British Library

ISBN: 978-0-12-374240-7

For information on all Academic Press publications
visit our web site at books.elsevier.com

Printed and bound in the USA
08 09 10   9 8 7 6 5 4 3 2 1

◼ CONTENTS

3 The Pathophysiology of Metabolic Syndrome

4 Metabolic Syndrome and Psychiatric Illness

5 Psychiatric Medications and Metabolic Syndrome

6 Depression, Metabolic Syndrome, and Heart Disease

7 Metabolic Syndrome, Insulin and Alzheimer's Disease

8 Metabolic Syndrome, Sleep, and Sex

9 Diets for Weight Loss and Metabolic Syndrome

10 Nutritional Supplements and Metabolic Syndrome

11 Conclusion: Metabolic Syndrome and what to do about it

ACKNOWLEDGMENTS

I wish to thank Jerry Reaven for his vision, intelligence, and generosity in sharing his story with me. Charlie Nemeroff is not only one of the most highly regarded biological psychiatrists in the world, but also one of the busiest. Thus, I am very grateful for his taking time to write the foreword to this book. Finally, I wish to express my love and thanks to my wife, Melissa, and my children, Ethan, Erin and Laurel. They sacrificed many a weekend to give me the freedom to write my book. My dears, we can play now.

■ FOREWORD

This remarkable volume by Scott D. Mendelson, M.D., Ph.D., will be a welcome addition to virtually every busy clinician's shelf. In fact, it probably will not reside on the bookshelf, but instead will sit on most practitioners' desks for easy availability to consult. There are a myriad of books published each year related to various aspects of diagnosis, neurobiology and treatment of one or another psychiatric disorders. But this volume is distinctive for several reasons. First and most important is the fact that there is increasing evidence that patients with psychiatric disorders, in particular schizophrenia, bipolar disorder and major depression, are at increased risk for a variety of maladies, all of which are contributed to by the so-called metabolic syndrome. Thus, the life expectancy of these patients is markedly reduced compared to the general population. This is likely because these disorders share a common pathophysiology with certain of the major psychiatric disorders. By that I mean, part of the underlying pathogenesis of cardiovascular disease likely also contributes to the pathogenesis of mood disorders. Patients with major psychiatric disorders must be viewed as a significant "at risk" population and, as such, need to be followed closely for signs and symptoms of the metabolic syndrome.

A second reason why this book is so important is because there is little doubt that psychiatric patients as a group receive less quality medical care than patients without psychiatric disorders. This is in part due to stigma, a battle that we are all still fighting, but it is also partly due to the difficulties that patients with cognitive and mood disturbances have in obtaining quality medical care. Recently in discussion with a prominent emergency room physician at an Ivy school in the northeast, he indicated to me his experience in which patients with "large psychiatric charts" simply do not receive the kind of quality care in the emergency room that other patients do. He agreed with my own experience that such patients coming into the emergency department with chest pain are more likely to be "worked up" for a myocardial infarction and less likely to be sent out of the emergency department for a follow-up visit at a clinic. They fall, unfortunately, into the category of patients memorialized in the "House of God", namely GOMERS (Get Out of My Emergency Room). Thus, their physical symptoms were often largely attributed to their psychiatric syndrome. How ironic this is when all of the available scientific evidence suggests that they are more, not less, likely to suffer from heart attacks, stroke and other consequences of the metabolic syndrome. A third major reason why this book is a welcome addition to our armamenterium is the fact that approximately a third or more of psychiatric patients have no other physician overseeing all of their medical care other than their psychiatrist.

Psychiatry in the 1950s moved away from the mainstream of medicine embracing purely psychoanalytic models of psychopathology and treatment. Many practitioners entered the field who were uncomfortable with mainstream medicine. Such clinicians felt it a virtual ethical violation to "lay hands on a patient" and as such, refused to measure their waist circumference, check their body weight or raise issues concerning the physical health that potentially could impact negatively on their psychotherapeutic treatment. In the last two decades, coined by Congress "the Decade of the Brain and the Decade of the Mind" respectively, psychiatry embraced evidence-based medicine and began the arduous process of determining which psychiatric treatments work and of those that did, how well did they work. This direction for the field coupled with the mounting evidence as described by Dr. Mendelson, in this concerning some common pathophysiologies between depression and disorders characterized by the metabolic syndrome such as diabetes, heart disease and stroke, have helped change the thinking of an entire new generation of psychiatrists. They view themselves as physicians who specialize in psychiatry, compared to the older generation who viewed themselves as psychiatrists, not physicians.

The fact that a sizeable percentage of depressed patients exhibit abnormalities in the immune system characterized by an increase in inflammatory cytokines, and that similar findings have been found in non-depressed patients with cardiovascular disease, is just one example of the commonalities between so-called medical syndromes associated with the metabolic syndrome and one of the major psychiatric syndromes. In the past several years, it has become clear that certain antipsychotic medications have the propensity to markedly increase body weight and adversely affect plasma lipid profiles. This is a particular problem for patients who require long-term treatment with these agents, such as those with severe bipolar disorder and schizophrenia. It has been quite surprising to me that the field took so long to recognize these problems and to respond to them. The agents that were the greatest offenders in terms of increased weight gain continued to be prescribed at an alarming rate, even when other equally effective atypical antipsychotics were available. This is a sad commentary on the lack of impact of evidence-based medicine on prescribing habits, and even more so on the influence of pharmaceutical company marketing in order to sway prescribing; only when a number of class action lawsuits were brought against drug companies that marketed these agents did prescribing habits finally change.

It is my hope that books such as this one will help prevent a recurrence of such poor prescribing practices. After reading this volume, why any practitioner would choose an agent associated with marked weight gain as a first choice for any given patient, surely would be a mystery.

Mendelson's description of the discovery of the metabolic syndrome by Gerald Reaven is a tour-de-force of both the history of science and personal discovery. His description of the metabolic syndrome is clear and he does a superb job in describing a very complex pathophysiology, taking care to distinguish between theory and fact in this evolving field. By reminding us at the end of this book to remember Hippocrates, Mendelson is really reminding us that we are physicians first and mental health care professionals second. These are important words to heed. If this advice is followed, the information contained in this volume, taken into consideration in practice, will reap tremendous benefits for our patients and for how psychiatry is viewed by our colleagues in other branches of medicine.

Charles B. Nemeroff, M.D., Ph.D.
Reunette W. Harris Professor and Chairman
Department of Psychiatry and Behavioral Sciences
Emory University School of Medicine
Atlanta, GA

1
GERALD REAVEN AND THE DISCOVERY OF SYNDROME X

For years, patients paraded before their doctors with obesity, high blood pressure, elevated cholesterol, heart disease, strokes, and diabetes. By and large, these doctors assumed that those conditions existed in the same patients simply because they were so unhealthy. In the 1960s, Dr. Gerald Reaven came to suspect that a common thread connected these seemingly separate conditions. He believed they were connected through the phenomenon of insulin resistance and to what he would later call Syndrome X. At that time his ideas flew in the face of the common understanding of the action of insulin. When I asked him how it was that he was able to see insulin resistance as the primary state underlying so many serious health problems when thousands of physicians and medical researchers did not, he answered with a disarming combination of humility and utter self-assuredness, "Because I believed the data, and they didn't."

My first interaction with Dr. Reaven came about when I wrote to him requesting copies of some of his scientific papers. I had expected either no answer at all or, at best, a brief form letter from his secretary with the requested papers enclosed. I was surprised when I received a cordial reply from Dr. Reaven himself, in which he sent me the papers and thanked me for my interest in his work. It was then that I considered the possibility of meeting with him. I was curious as to how he conceived of insulin resistance and how the notion evolved into his formulation of Syndrome X, which is now more commonly referred to as Metabolic Syndrome. I particularly hoped to discover what it was in his nature that drove him to persevere, often under scathing criticism, in a line of research that would eventually lead him to one of the most important discoveries in the last quarter-century of medical research. Reaven generously agreed to meet with me, and over dinner I learned a great deal about him. He gave me a spirited history of his intellectual life and how he came to consider, and later firmly establish, the existence of Syndrome X.

Jerry Reaven grew up without the slightest thought of becoming a physician. "There was no way in hell I was going to be a doctor," were his precise words. "I was a ball player," he added with a grin. When I asked what kind of ball, he replied as if it was a given, "All kinds." However, a life-changing event occurred in 1944 when he was a sophomore in high school. He underwent a physical examination that was required to join the school basketball team. It included a routine skin test for tuberculosis, and much to his shock he tested positive. He was sent for treatment at a sanitarium. In retrospect he believes that he did not actually have tuberculosis. Nonetheless, his dreams of playing ball ended there.

While at the sanitarium a physician took an interest in him. He encouraged Jerry to apply for entrance in a new undergraduate program at the University of Chicago. Because they encouraged early admission into college studies, they were willing to accept him straight out of his uncompleted sophomore year of high school. The program had been designed by the illustrious, albeit controversial, American educator Robert Hutchins. The program was based on the study of the classic books of western civilization. Hutchins himself had once remarked that the great books were the ideal basis for a liberal education, "if only because they were teacher proof." The students consumed volumes of those great works. They pondered and debated ideas. They challenged their professors and each other. "There was no respect for authority," Reaven said with a wry look. "The professors had to earn it." He looks back on those days with great affection, and it was to his undergraduate education at the University of Chicago that he attributes his ability to question the prevailing notions and follow the thread of truth wherever it led.

After graduating, he intended to pursue graduate work in English literature. However, his family saw no financial security in it. In uncharacteristic acquiescence to family pressure, he agreed to attend medical school as two of his uncles had done. He stayed at the University of Chicago and attended their medical school. He hated it. It was the antithesis of the style of learning he had come to love in his undergraduate training. In medical school there was no questioning or debating. Its purpose seemed to be to fill you up with the accepted medical dogma of the day. He despised the boring lectures. Instead of studying his textbooks, he would often spend hours in the library pursuing answers to questions that may have occurred to him during the day. He recalls that he "barely made it".

At one point during medical school, he had considered becoming a psychiatrist. The brain and how it controlled behavior fascinated him. However, he had distaste for the psychoanalytic approach to psychiatry that was in vogue at the time. In his first psychiatry clerkship he fully realized that "talk therapy" was not for him. He admitted with a smile that he just did not like having to listen to somebody else for so long. He had wanted to pursue what is now generally referred to as Biological Psychiatry. Unfortunately, he was ahead of his time. There were no programs for psychiatrists based in neuroscience, and the field of psychopharmacology had not yet even come into being. Perhaps it was this interest about the brain from years ago that I tapped into when I first discussed the book I planned to write about Metabolic Syndrome from a psychiatrist's perspective.

It was not until his last year of medical school, when he began his clerkships in Internal Medicine, that a passion for medicine took hold. The professors of Internal Medicine devoted most of their time to medical research. They encouraged discussions of cases and the relevant science. It was again like his undergraduate days at the University of Chicago. "We discussed what was known about an illness and what was unknown. We talked about how to find out those things we didn't know. I loved it!"

At that time at the University of Chicago there was a brilliant urologist and surgeon, Dr. Charles Huggins. Dr. Huggins was making tremendous strides in the treatment of prostate cancer. He had found that cancers of the prostate were extremely sensitive to testosterone in the blood, and that lowering serum testosterone could prolong patients' lives. In 1966, he was awarded the Nobel Prize for his work. While Jerry was in his first year of residency, and going through the rotations that were required of interns, he started work on the urological service. Dr. Huggins took a liking to him. He began to groom Jerry to become a urological surgeon. However, in what must have been a great horror to Huggins, Jerry found himself fascinated with aspects of urology that most self-respecting surgeons despised. He was more interested in balancing the salt metabolism and fluids of patients at their bedsides than in brandishing a scapel above them in the operating theater. Dr. Gerald Reaven was not to become a surgeon.

With his growing interest in fluids and electrolyte imbalances, Jerry was drifting into what was then the realm of the endocrinologists. He decided to move to the University of California at San Francisco (UCSF) to work with Dr. John Luetscher. Luetscher was a renowned endocrinologist and a codiscoverer of the hormone aldosterone. Aldosterone had been found to be a major factor in the control of serum sodium, fluid, blood pressure, and kidney function. It was essential to life, but when excreted in excess it contributed to hypertension and heart disease. It was an exciting area of study. Unfortunately, Jerry wasn't happy with the academic atmosphere at UCSF, and after a year or so he decided to work elsewhere.

He then traveled to the University of Michigan to start work with another famous endocrinologist, Dr. Jerome Conn. Conn was, as Jerry likes to say, "a hot rock." Like Luetscher at UCSF, Conn was performing research in the pathophysiology of adrenal function. Conn's Syndrome, a disease of excessive aldosterone secretion, bears his name. However, Jerry did not really consider himself an endocrinologist at Michigan. He explained that at that time endocrinologists and nephrologists were lumped together at academic medical centers. With his interest in salt and fluid balance, he felt more kinship with the nephrologists. Moreover, he admitted that he was harboring a growing distaste for endocrinologists, particularly those that studied and treated diabetes.

"I had met some diabetologists who were really stupid!" he said unabashedly. "It was obvious that what they were teaching was wrong. They were saying that there was only one disease called diabetes. If you were young and didn't have insulin, they called it juvenile onset diabetes. If you were old with high serum glucose they called it the maturity onset form of diabetes. But it was obvious as hell that they weren't the same disease! When you were giving the kids a little insulin, you always had to worry about hypoglycemia, but some of the older people you could give tons of insulin. I felt I knew more about the disease than the diabetologists did." It was amusing to hear the man who 30 years later would be awarded the Banting Prize, the world's most prestigious award for research in diabetes, tell me, "I had no interest in diabetes whatsoever."

Almost from the beginning Jerry had misgivings about working in Jerome Conn's laboratory. "He was a very bright guy, but he ran the place the old-fashioned way. He was the general, he had colonels, and it was clear pretty early on that I wasn't going to be a good first lieutenant. I'm too independent." "Everyone liked me at Michigan," he said with a shoulder shrug of innocence, "but I suspected they weren't going to like me very long. I decided I would look elsewhere." He wrote to his old mentor, John Luetscher, who by that time had relocated at Stanford's new medical school in Palo Alto, California. Luetscher convinced him to return to California and seek a position at Stanford. Reaven joined the faculty of the Stanford medical school soon after it opened in 1959, and he has remained there ever since.

Reaven became more interested in diabetes in 1960 after a paper by Drs. Rosalyn Yalow and Solomon Berson blew the roof off diabetes research.[1] Yalow and Berson developed a new biochemical technique that allowed them to quickly and definitively measure the amount of insulin in human blood. Their new technique, called radioimmunoassay, was so important a scientific breakthrough that Yalow received the Nobel for Physiology and Medicine in 1977. Of interest to Jerry Reaven and the rest of the medical community was that in evaluating insulin levels in normal and diabetic patients, Yalow and Berson found that many individuals with maturity onset diabetes were not deficient in insulin. In fact, some of those patients diagnosed as having diabetes actually had elevated levels of insulin in their blood. For Reaven, these findings confirmed his suspicions that juvenile diabetes and maturity onset diabetes were two distinct illnesses. In the first case, the patients had no insulin, whereas in the second case, they had insulin, but for some reason the insulin did not work. It occurred to him that the body must develop some kind of resistance to the effects of insulin.

Soon after the 1960 paper by Yalow and Berson, other research groups, including Reaven's group at Stanford, began to publish studies confirming that individuals with maturity onset diabetes could have normal or even high levels of insulin. These data strengthened the likelihood that some people produced sufficient amounts of insulin but were resistant to it. Nonetheless, many diabetologists held fast to the notion that people with maturity onset diabetes, like those with juvenile onset diabetes, had an insufficient supply of insulin. They argued that the pancreas might produce insulin, but it was not always released in adequate amounts. Others suggested that it was not released at the right time. They refused to believe the body could suffer from ineffectiveness of insulin in the midst of plenty of the hormone.

In the mid-1960s, Reaven began work on a method that would give unassailable evidence of insulin resistance in maturity onset diabetes. Their first formal results were published in 1970[2]. Subjects in the study, some normal and some with maturity onset diabetes, were intravenously administered the same amount of insulin over several hours. At the same time, they were given small amounts of adrenaline to eliminate release of insulin from their own pancreas. Propranolol was also given to minimize any release of glucose by the liver. With those measures in place, the subjects were given the same amount of glucose intravenously. After a time, Reaven's group measured the serum glucose and insulin levels in all the subjects. The results were dramatic. Although the subjects with maturity onset diabetes had the same amount of insulin in their blood, they had substantially higher levels of glucose than did the normal subjects. Reaven and his coworkers thus showed conclusively that people with maturity onset diabetes could have hyperglycemia despite normal levels of insulin. They were insulin resistant. Despite this compelling evidence, it took another 12 years before insulin resistance became an accepted part of medical thought.

Reaven's work in developing a method to precisely measure insulin resistance was a breakthrough. However, even as he was helping establish the existence of insulin resistance, he was already looking beyond it to the bigger picture. He was particularly interested in exploring what he suspected to be links between insulin, carbohydrate, fat, and heart disease. Reaven explained that from the establishment of the phenomenon of insulin resistance, as well as from results of several landmark papers from the previous years, he began to build a hypothesis about how insulin resistance might increase the risk of heart disease.

Reaven noted that in 1958, Margaret Albrink and Evelyn Man from the Yale University Medical School had published a paper showing that hypertriglyceridemia was at least as important as high cholesterol in predisposing individuals to coronary artery disease[3]. The relationship between triglycerides and carbohydrate in the diet began to emerge from the experiments of Dr. Edward Ahrens at the Rockefeller University. Dr. Ahrens and a group that included John Farqhuar, who was later to work with Jerry Reaven, showed that most people with hypertriglyceridemia acquired that condition not from eating too much fat, but rather from consuming too much carbohydrate[4]. The link between poor glucose control, hypertriglyceridemia and heart attack became clearer after a 1963 study by Reaven and his coworkers at Stanford[5]. In that study Reaven discovered that patients who had recently suffered heart attacks not only had hypertriglyceridemia, but also had elevated serum glucose levels and inability to normalize those glucose levels after a meal of carbohydrate. That is, they were glucose intolerant. Reaven saw similarities between the high serum glucose levels in those patients and the high carbohydrate intake of Ahren's subjects with hypertriglyceridemia. In both situations, insulin levels would be high in the body's attempt to lower serum glucose. He suspected that insulin resistance and compensatory high levels of insulin could be factors in causing hypertriglyceridemia and, in turn, heart disease.

In 1967, Reaven nailed down a major link between increased levels of insulin, hypertriglyceridemia, and increased risk for heart disease[6]. He had previously determined that insulin stimulates the liver to produce triglyceride, which Albrink and Man had found to be a major contributor to heart disease. In this study he showed that in patients whose muscle and adipose tissue had become resistant to the effects of insulin, the liver still dutifully responded to the insulin. While the pancreas pumped out large amounts of insulin to coax resistant tissues to take up glucose, the ever-increasing levels of insulin prompted the liver to pour out what was in some cases extremely high levels of triglyceride.

High levels of the cholesterol-bearing proteins Very Low Density Lipoprotein (VLDL) and Low Density Lipoprotein (LDL) are associated with increased risk of cardiovascular disease. When the liver is stimulated to produce triglyceride, it cannot help but produce VLDL along with it because VLDL is how the liver packages triglyceride for release into the blood. When the liver gets revved up and produces too much triglyceride, the patient not only suffers hypertriglyceridemia, but from high serum levels of VLDL as well. Enzymes in the body can subsequently convert VLDL into LDL, thus the process also tends to cause high serum levels of that more dangerous form of cholesterol.

As early as 1966, Reaven and his coworkers were showing that insulin resistance and the compensatory increases in serum insulin after meals were associated with both hypertriglyceridemia and high serum VLDL[7]. However, in the 1970s, things became more complicated, when evidence began to show that one form of cholesterol, High Density Lipoprotein (HDL), might help prevent cardiovascular disease. In fact, it appeared that low levels of HDL could contribute to heart disease as much as high levels of LDL[8]. Reaven then hypothesized that increases in serum insulin as a compensatory effect secondary to insulin resistance, might also be responsible for low levels of HDL in certain individuals.

In 1983 Reaven reported that in normal subjects fed diets of 60 percent carbohydrate, serum triglycerides increased, whereas levels of HDL decreased[9]. In the subjects fed the high carbohydrate diet, insulin levels were significantly increased after meals. He concluded that the increased levels of insulin caused decreases in HDL, as well as increases in both triglyceride and VLDL. At that time, scientists had only begun to understand the complicated process of transfer of cholesterol from HDL to VLDL by cholesteryl ester transfer protein[10]. However, Reaven was well aware of evidence that interaction between HDL and VLDL could deplete the good form of cholesterol. For him the picture began to more clearly suggest that low HDL, hypertriglyceridemia and high VLDL could all occur as the result of insulin resistance.

There has long been known to be a relationship between diabetes and hypertension. In the mid-1980s Reaven turned his attention toward finding a possible relationship between hypertension and insulin resistance. Dr. Reaven and his coworkers discovered that many patients with untreated hypertension had high serum levels of insulin and glucose intolerance. These same patients also tended to have low HDL and high LDL[11]. He further discovered that patients treated for hypertension with medication continued to show insulin resistance even after their blood pressures had normalized. It was concluded that hypertension did not cause the resistance to insulin; rather, insulin resistance might be at least partially responsible for the hypertension[12].

Thus, hypertension began to emerge as yet another major risk factor for cardiovascular disease that could be related to insulin resistance.

Over 25 years, Jerry Reaven and his team collected scientific data concerning insulin resistance, carbohydrate intake, serum glucose, triglycerides, cholesterol, diabetes, and heart disease. In many cases he invented technology to tease out and measure effects of insulin on the liver, muscle, and fat tissues. He told me that as early as the mid-1960s, he had become completely convinced that insulin resistance could emerge in certain individuals, and that it increased the risk for diabetes and heart disease by increasing serum glucose, insulin, cholesterol and triglycerides. He slowly and methodically built his case.

Although Reaven's work is now almost universally accepted, his ideas were initially met with considerable skepticism. He recalls in his book "Syndrome X", that after his first reports about insulin resistance, "The scientific world did not stand up and applaud." He noted a time when his colleague, John Farquhar, lamented the fact that their work had gained so little recognition and acceptance from the medical community. However, Reaven, himself, was philosophical about it. "My research made me a better clinician and a better teacher to my students," he explained. "You keep doing what you're doing. If you think you're right, you keep saying it, and that's it. Our task wasn't to convince the world, it was to do good work!"

At times he was the target of outright hostility. As we spoke, he recalled a meeting he attended in New Orleans in the late 1960s. A group of diabetologists who had given themselves the quaint name, "The Southern Sugar Club" had invited him to deliver a lecture on his new theory of insulin resistance. After he spoke, one of the leading members of the club marched to the lectern and proceeded to berate him. She informed him and the rest of the audience that the world authorities had met the previous month in Stockholm, Sweden and had concluded that insulin resistance simply did not exist! "Why do you keep talking about it?" she asked in the rhetorical sense. "Jerry," she told him, "just give it up!" Ironically, it was 20 years later in New Orleans that he was awarded the Banting Prize, largely due to his establishment of insulin resistance as a major cause of human disease.

In 1988, in his lecture at the Banting Prize award ceremony, Dr. Reaven formally presented his theory of insulin resistance and its pathophysiological sequelae[13]. He stated that genetic predisposition and dietary indiscretion could lead to insulin resistance. Individuals who are resistant to insulin require abnormally high levels of insulin to maintain a ceiling on serum glucose levels. These patients have high fasting insulin levels, poor glucose tolerance, hypertension, low serum HDL, high VLDL, and hypertriglyceridemia. He called this constellation of problems "Syndrome X", deriving the "X" from his recollections of high school algebra, where "X" was the unknown factor. Insulin resistance was the previously "unknown factor" that robbed so many of their health. It was the primary factor in Syndrome X.

Reaven went on to explain in his Banting address that Syndrome X led a patient down one of two roads of pathology. As long as the beta cells of the pancreas could produce enough insulin, serum glucose would stay within fairly normal limits even in patients with insulin resistance. However, the price one paid for the high levels of insulin necessary to control serum glucose was an increased risk of cardiovascular disease due to hypertriglyceridemia, increased levels of VLDL and LDL, decrease in HDL, and hypertension. On the other hand, if the beta cells could not keep up with ever increasing demand for insulin in Syndrome X, then an uncontrolled rise in serum glucose levels would eventually herald the onset of Diabetes Type II.

In the years following his proposal of Syndrome X, Reaven and many others have confirmed and expanded our understanding of the pathophysiology of insulin resistance. The list of abnormalities that occur as a consequence of chronic insulin resistance has also grown, and the "X" that named his syndrome is no longer an unknown factor. The syndrome is now referred to as Metabolic Syndrome.

A relationship that has become increasingly important since the 1988 Banting lecture is that between Metabolic Syndrome and obesity. Reaven was aware that individuals with obesity tend to have many of the pathologies that have been associated with insulin resistance. He stated quite clearly in his lecture that avoiding obesity and remaining physically active helps reduce the risk of subsequently developing insulin resistance, diabetes, and cardiovascular disease. However, he emphasized the fact that many obese individuals have perfectly adequate responses to insulin. Moreover, some patients that do not meet standard criteria for diagnosis of obesity may still have significant insulin resistance, hypertension and hyperlipidemia. Thus, Reaven initially had no compelling basis to see obesity *per se* as an integral part of his Syndrome X.

Since the 1990s there has been growing awareness of different types of obesity. Of particular interest have been studies showing that the incidence of Metabolic Syndrome and risk of developing cardiovascular disease is increased in people that accumulate disproportionately large amounts of fat within their abdominal cavities. It has recently become apparent that the adipocytes that populate visceral fat are different from those in other parts of the body. The visceral adipocytes are extraordinarily complex and secrete a plethora of hormonally active substances referred to as adipocytokines. The presence or lack of certain adipocytokines can adversely affect the body's sensitivity to insulin and contribute to Metabolic Syndrome. In some cases, adipocytokines can independently produce some of the adverse effects seen in the syndrome. Consequently, the World Health Organization and The United States National Institute of Health have included abdominal obesity in their lists of defining characteristics of Metabolic Syndrome. The current list of the five defining characteristics of Metabolic Syndrome, according to the National Cholesterol Education Program's Third Adult Treatment Panel includes: waist circumference at the umbilicus over 102 cm in men and 88 cm in women; fasting serum glucose over 110 mg/dl; serum triglycerides above 150 mg/dl; HDL levels below 40 mg/dl in men and 50 mg/dl in women; and blood pressures over 130 mm Hg systolic or 85 mm Hg diastolic. If three or more of those abnormalities are found, a patient meets criteria for diagnoses of Metabolic Syndrome[14].

Ironically, Reaven himself has recently come to see the concepts of both Syndrome X and Metabolic Syndrome as inadequate and overly simplistic. He has since described what he refers to as the Insulin Resistance Syndrome[15]. The Insulin Resistance Syndrome includes the originally noted insulin resistance, hyperinsulinemia, hyperglycemia, high blood pressure, hypertriglyceridemia, and low HDL. However, in addition, Reaven has come to understand that inflammation, abnormalities of blood clotting, dysfunction of arterial endothelium, abnormalities of uric acid metabolism, and various types of tissue hypertrophy can also occur in chronic states of insulin resistance. Along with diabetes and cardiovascular disease, he now includes conditions as diverse as polycystic ovary disease, sleep apnea, gout, nonalcoholic fatty liver disease, and certain forms of cancer as consequences that can arise from insulin resistance.

A subject that has not been adequately explored is the relationship between Metabolic Syndrome and psychiatric illness. In fact, the two interact and often exacerbate each other through processes of inflammation, stress, sleep disturbance, autonomic hyperactivation, and other factors, in addition to the adverse effects of insulin resistance and compensatory hyperinsulinemia. Psychiatrists and others who treat patients with mental illness have tended to associate Metabolic Syndrome and psychiatric illness only in terms of medication-induced weight gain, hyperlipidemia, and diabetes. Certainly, these problems are serious and deserve attention. However, such narrow focus obscures the fact that the relationships between psychiatric illness and Metabolic Syndrome are far deeper and more complex.

REFERENCES

1. Yalow, R.S. and Berson, S.A., Immunoassay of endogenous plasma insulin in man. *J. Clin. Invest.* 1960; 39:1157–1167.
2. Shen, S.-W. et al., Comparison of impedance to insulin-mediated glucose uptake in normal subjects and in subjects with latent diabetes. *J. Clin. Invest.* 1970; 49:2151–2160.
3. Albrink, M.J. and Man, E.B., Serum triglycerides in coronary artery disease. *Trans. Assoc. Am. Phys.* 1958; 71:162–173.
4. Ahrens, E.H. et al., Carbohydrate-induced and fat induced-induced lipemia. *Trans. Assoc. Am. Phys.* 1961; 74:134–146.
5. Reaven, G.M. et al., Carbohydrate intolerance and hyperlipemia in patients with myocarduial infarction without known diabetes mellitus. *J. Clin. Endocrinol. Metab.* 1963; 23:1013–1023.
6. Reaven, G.M. et al., Role of insulin in endogenous hypertriglyceridemia. *J. Clin. Invest.* 1967; 46:1756–1767.
7. Tobey, T.A. et al., Relationship between insulin resistance, insulin secretion, very low density lipoprotein kinetics, and plasma triglyceride levels in normotriglyceridemic man. *Metabolism* 1981; 30:165–171.
8. Miller, N.E. et al., The Tromso Heart Study. High density lipoprotein and coronary heart disease: a prospective case-control study. *Lancet* 1977; I:965–968.
9. Coulston, A.M., Liu, G.C. and Reaven, G.M., Plasma glucose, insulin and lipid responses to high-carbohydrate low-fat diets in normal humans. *Metabolism* 1983; 32:52–56.
10. Pattnaik, N.M. et al., Cholesteryl ester exchange protein in human plasma. Isolation and characterization. *Biochem. Biophys. Acta* 1978; 530:428–438.
11. Fuh, M.M. et al., Abnormalities of carbohydrate and lipid metabolism in patients with hypertension. *Arch. Intern. Med.* 1987; 147:1035–1038.
12. Shen, D.C. et al., Resistance to insulin-stimulated-glucose uptake in patients with hypertension. *J. Clin. Endocrinol. Metab.* 1988; 66:580–583.
13. Reaven, G.M., Role of insulin resistance in human disease. *Diabetes* 1988; 37:1595–1607.
14. National Cholesterol Education Program (NCEP) Expert Panel on Detection, Evaluation, and Treatment of High Blood Cholesterol in Adults (Adult Treatment Panel III). Third report of the NCEP panel on detection, evaluation and treatment of high blood cholesterol in adults (adult treatment panel III). *Circulation* 2002; 106:3143–3421.
15. Reaven, G.M., Why Syndrome X? From Harold Himsworth to the Insulin Resistance Syndrome. *Cell. Metab.* 2005; 1:9–14.

2
FACTORS THAT CONTRIBUTE TO METABOLIC SYNDROME

It is likely that many indiscretions of modern diet and lifestyle eventually lead to Metabolic Syndrome. Some people are genetically predisposed to it. However, various combinations of too much sugar, too much saturated fat, too many calories, too little exercise, too little sleep, and too much stress can initiate and accelerate the pathological processes. Metabolic Syndrome can be further aggravated by medical and psychiatric comorbidities, and, in some cases, by adverse effects of medications used to treat them. The prevalence continues to grow. A recent report from the US Center for Disease Control estimated that as many as 35 percent of men and women in the United States meet the criteria for Metabolic Syndrome.

GENETICS

Unfortunately for many individuals, there is a strong genetic component to Metabolic Syndrome. Although Metabolic Syndrome is largely due to insulin resistance, sensitivity to insulin can vary even among people who appear to be perfectly healthy. The insulin sensitivity of Pima Indians, who are prone to developing diabetes, was compared with the sensitivity of subjects of European descent. Marked differences were noted between the two groups in the efficiency with which insulin lowered serum glucose levels. After differences in weight and fitness were taken into account, it was concluded that up to 50 percent of the differences were due to genetic factors[1]. Genetic factors are also evident in the fact that members of families tend to be similar in their

Metabolic Syndrome and Psychiatric Illness: Interactions, Pathophysiology, Assessment & Treatment

responses to insulin[2]. Similarities in response to insulin are stronger between identical twins than fraternal twins[3].

Before Dr. Reaven described Syndrome X, Dr. James Neel proposed the "thrifty gene" hypothesis[4]. Neel's theory may provide an evolutionary basis for genetic predisposition to Metabolic Syndrome. Neel speculated that early man suffered periods of scarcity of food and starvation. Consequently, it was much to his advantage to have a metabolism in which calories could be efficiently packed away when food was plentiful. When food was unavailable, the body could be frugal with the calories it had stored away. Neel's theory predicted that animals prone to obesity could withstand periods of scarcity better than animals not able to pile on the pounds when food is available. Experiments have born out that mice genetically prone to obesity are better able to withstand long periods of food deprivation than normal mice[5]. Neel went on to speculate that while thrifty genes are useful during food shortage, during periods of plenty, such as we in modern Western society enjoy, these genes can work against us. He suspected that enhanced ability to store calories during persistent states of plenty was a major cause of obesity and diabetes.

Medical scientists are starting to identify genes that predispose people to Metabolic Syndrome[6]. They tend to be ones that would be expected to do so by Neel's thrifty gene hypothesis. Most of the genes under scrutiny are ones that code for proteins involved in the control of energy metabolism. These proteins include CD36, which controls the uptake of fat into cells, and PPAR, which controls the metabolism of fat within cells of the body. The enzyme 11-beta-hydroxysteroid dehydrogenase, which modulates levels of the stress hormone cortisol, and beta adrenergic type 3 receptors that mediate effects of the autonomic nervous system system on energy metabolism have been implicated. Also suspected are receptors for melanocortins, which play a role in integrating mammalian metabolism with seasonal changes in food supply[7], and the 5-HT2C receptor. Mutations of the 5-HT2C receptor have been found to predispose both laboratory animals and humans to obesity[8]. Mutations of melanocortin[9] and 5-HT2C[10] receptors have also recently been implicated in affective disorders.

CARBOHYDRATES AND THE GLYCEMIC INDEX

Although genetic background is important in determining who may develop Metabolic Syndrome, at least half of the risk for developing Metabolic Syndrome arises from diet and other lifestyle choices. Many authorities, including Reaven, suggest that reduced consumption of carbohydrates is a prudent step to take in the prevention and treatment of Metabolic Syndrome. The logic of this dietary strategy is fairly straightforward. When carbohydrates supply a significant proportion of calories taken in, more insulin needs to be released into the blood to control the resulting increases in serum glucose. In individuals who are resistant to insulin, abnormally high levels of insulin must be pumped out to compensate for the resistance. This hyperinsulism stimulates the liver to produce and release more triglyceride, which, in turn, contributes to increases to insulin resistance in muscle and adipose tissues.

A factor that has become increasingly recognized as important is the Glycemic Index. Different sources of starch and sugar can produce different levels of serum glucose after ingestion. A table of comparisons among carbohydrate sources, called The Glycemic Index, was conceived by Dr. David J.A. Jenkins in 1981[11]. Foods were all compared with a slice of white bread, which was given the arbitrary glycemic value of 100. The glycemic value of a food is not always what intuition might suggest. A number of factors, such as the presence of fat and fiber in food, can reduce the rate at which a carbohydrate is digested and absorbed. Tables of the Glycemic Index values for foods are available in bookstores and on the Internet. However, to illustrate the fact that not all is as it may seem, I note that while the glycemic value of white bread is 100, the value of rye bread is 58, baked potato is 135, lentil is 43, pure glucose is 138, table sugar is 86, fructose is 30, ice cream is 52, raisin is 93, and plum is 34.

Carbohydrates with high glycemic index are quickly digested and rapidly raise serum glucose. On the other hand, carbohydrates with low glycemic index are slowly digested and absorbed. There are advantages to foods that only slowly increase serum glucose, as they do not stimulate as much insulin release. Insulin resistance has been found to increase as the intake of high glycemic index carbohydrates in the diet increases[12]. Bursts of insulin can also precipitate abnormally low levels of glucose, that is hypoglycemia. Hypoglycemia generates stress, increases hunger, causes carbohydrate craving, and initites a vicious cycle of poor food choices.

Due to genetic variation, it is likely that some individuals can do perfectly well on high carbohydrate diets. This is particularly the case if the carbohydrates include whole grains and vegetables that are high in fiber and low in the Glycemic Index. An interesting question is whether a person who is not genetically predisposed to insulin resistance can become resistant by imprudent intake of high glycemic index carbohydrates. Many, if not most, people will show an increase in serum triglycerides when challenged with large amounts of carbohydrate. If prolonged in nature, this could exacerbate an existing or evolving state of insulin resistance and Metabolic Syndrome. Moreover, regardless of genetic variation, it is a basic part of human physiology for extra calories from high-carbohydrate diets to be converted into the saturated fatty acid, palmitic acid[13]. Palmitic acid is well known for its ability to increase levels of cholesterol and promote fat deposition in coronary arteries and other tissues of the body. High serum levels of palimitic acid are associated with increased risk of Metabolic Syndrome[14]. There may also be relationships between the likelihood of carbohydrate-induced hypertriglyceridemia and factors such as sex, obesity, and baseline triglycerides. Unfortunately, there is no certain way to predict who is likely to show increases in triglycerides while consuming a high-carbohydrate diet[15]. Nonetheless, existing evidence suggests that *any* individual will begin to develop signs of Metabolic Syndrome if high-glycemic index carbohydrates are given in high enough quantities for long enough periods of time. A common technique to produce Metabolic Syndrome in otherwise healthy laboratory rats is to keep them on a steady diet of large quantities of sucrose or fructose.

FRUCTOSE

One of the most widespread, dangerous, and under-appreciated dietary contributors to Metabolic Syndrome is fructose. Fructose is a natural sugar. It has a low glycemic index and it does not stimulate the release of insulin. For a time it was seen as a useful substance for the treatment of reactive hypoglycemia[16], or as a source of carbohydrate for athletes who needed a quick burst of energy without risk of hypoglycemia[17]. There is reason to believe that small amounts of fructose can be helpful in the treatment of diabetes. Unfortunately, intake of large quantities of fructose is a cause of serious metabolic dysfunction in the body.

While consumption of table sugar decreased over the last 25 years, yearly consumption of fructose has increased over this period of time from 0.5 lb to over 60 lb per person[18]. This increase is primarily due to an explosion in the use of high-fructose corn syrup. Fructose is sweeter than either sucrose or glucose, thus using enzymes to convert the glucose of cornstarch into sweeter fructose is economically advantageous. Because fructose, in the form of high-fructose corn syrup, is cheap and very sweet, it has become the darling of the food industry. Fructose is everywhere! It appears in soft drinks, baked goods, canned soups, pancake syrup, yogurt, ketchup, sauces, and a multitude of other processed foods.

Fructose and glucose are handled differently by the liver. The physiology involved is complex. However, it is safe, if overly simplistic, to say that the liver evolved primarily to process glucose. Too much or too little glucose trips biochemical control mechanisms that work to restore a balance between the supply and demand for energy. When fructose is metabolized, the major rate-controlling steps are bypassed. Fructose is an "unregulated" source of calories. High intake of fructose dramatically increases the production of fat by the liver, and the levels of triglycerides and free fatty acids in the blood[19]. Fructose does not stimulate release of either insulin or leptin, which work together in the brain to reduce appetite and produce feelings of satiety. On the other hand, fructose fails to reduce levels of the hormone ghrelin that ordinarily decreases with carbohydrate intake. Unlike insulin and leptin, ghrelin tends to stimulate appetite[20]. Thus, if an individual replaces sucrose with fructose in their meals, they are more likely to keep eating and consume more calories.

FIBER

Although fiber is technically a form of carbohydrate, it can also be seen as a distinct type of macronutrient. Many people think of fiber only in terms of relieving bowel irregularity. Certainly, fiber serves a very important role in this regard. However, an equally important effect of fiber is the way it changes absorption of other macronutrients. In particular, fiber slows down the absorption of carbohydrate, and acts to lower the glycemic index of high-carbohydrate foods. High fiber content in the diet lowers insulin resistance and improves serum lipids[21]. Fiber also slows the rate that the stomach empties and helps give a feeling of being full. Thus, high fiber content can help reduce the number of calories consumed in each meal, because people simply do not feel as hungry.

FATTY ACIDS AND HOW THEY DIFFER

The type and quantity of fats consumed are important in determining if Metabolic Syndrome will evolve over time. The types of fats in the diet include saturated, monounsaturated, polyunsaturated, and *trans* fats. Some are essential, some are healthy, some are deleterious, and some are outright deadly.

The terminology used to describe fatty acids is based on structural features in their carbon chains. For those not well-versed in organic chemistry, I will explain. The configuration of the electron shell of the carbon atom predisposes it to make four chemical bonds with other atoms. With the exception of those on the ends of the chain, each carbon in a fatty acid binds to two other carbon atoms, one on each side. This leaves two more places on each carbon for other atoms to bind. Hydrogen atoms can bind in the remaining spots, and when every remaining spot in the carbon chain is bound with hydrogen, the fatty acid is "saturated" with hydrogen. Hence the term saturated fat. In some fatty acids, carbons fill two binding slots with a single adjacent carbon, producing a double bound between them. There are fewer hydrogen atoms binding to the carbons in the double bound, and thus the fatty acid is "unsaturated." Fatty acids with one double bond are monounsaturated. Those with more than one double bond are polyunsaturated.

The single bonds between two carbon atoms allow the carbons to twist freely, like twisting a ball on the end of a stick. However, when there is a double bond, the carbon atoms are locked into place in relation to each other. There are two configurations into which double bounds can lock carbon atoms in the middle of a fatty acid chain. In one form, called *cis*, the ends of the carbon chain extend out on the same side of the plane defined by the double bond. The chain of carbons takes the shape of a "U," with the double bond at the bottom. The *cis* form of fatty acids is what occurs naturally in plants and animals. The other form of unsaturated fatty acid is the *trans* form. In this case, the ends of the chain advance out from the opposite sides of the plane defined by the double bound. The carbon chain looks something like an "N", with the double bond being the diagonal line.

With rare exceptions, the *trans* forms of fatty acids are not natural. They are the result of humans tinkering with fatty acids to give them specific qualities. Food manufacturers prefer the thick, creamy quality of saturated fats to the more oily unsaturated fats. The hydrogenation process that turns unsaturated into saturated fatty acids randomly breaks but often reestablishes carbon double bonds. Hydrogenation unavoidably introduces *trans* fatty acids into the mix, and it is the major culprit in increasing the quantity of the unnatural *trans* fatty acids in our food.

Fatty acids may also differ in terms of where a double bond is situated in relation to the *omega* carbon, which is the carbon on the end of the fatty acid opposite that of the acid group. Fatty acids with a double bond between the third and fourth carbon from the end are *omega-3* fatty acids. Fatty acids with double bonds between the sixth and seventh carbon from *omega* are *omega-6* fatty acids. While both types of fatty acids are required for health, the *omega-3* and *omega-6* are quite different from one another. They each have their role, and their intake needs to be balanced.

The structural differences among the fatty acids can make extraordinary differences in human health. Some fatty acids are like vitamins. That is, they are necessary for human health, but are not made in the body. These are the essential fatty acids, and they must be obtained in the diet. They include the *omega-3* alpha-linolenic acid and the *omega-6* linoleic acid. All the other fatty acids the body needs can be produced either from scratch or from those two essential fatty acids.

SATURATED FAT

The fatty acids the body makes to store extra energy from unused carbohydrate are saturated. This is also the case in lower animals we use as food. Thus, saturated fat is the major form we ingest when we eat meat. The human body can also produce monounsaturated fatty acid. We are able to turn about 12 percent of ingested stearic acid, one of the primary saturated fatty acids in beef, into oleic acid, the major monounsaturated fatty acid in olive oil. This has led some authorities to suspect that moderate amounts of beef and meats are not as risky as might be suggested by their saturated fat content. In fact, a study reported in *The New England Journal of Medicine* found that stearic acid is as effective as oleic acid of olive oil in improving serum cholesterol profiles[22].

Some saturated fats are clearly more unhealthy than stearic acid. For example, increased serum levels of palmitic acid, a saturated fatty acid found in tropical oils, is associated with increased risk of Metabolic Syndrome[23]. For reasons that are unclear, high intake of saturated fats also increases deposition of fat in the abdomen[24]. This pattern of fat deposition predisposes to Metabolic Syndrome.

The monounsaturated fatty acid, oleic acid can protect from Metabolic Syndrome. Switching from saturated to mostly monounsaturated fatty acid intake, as would occur in replacing butter with olive oil, both increases insulin sensitivity[25] and decreases LDL[26]. Olive oil, which contains the monounsaturated oleic acid, is a major component of the so-called Mediterranean diet. As I will discuss in a later section, this type of diet has been shown to prevent and reverse components of Metabolic Syndrome.

OMEGA-3 AND OMEGA-6

The ratio of *omega-3* to *omega-6* polyunsaturated fatty acids in our diet has changed through the centuries. While modern diets have *omega-6* to *omega-3* ratios as high as 20 to 1, ratios closer to 2 to 1 or lower may be optimal[27]. Ironically, the inclusion in our diet of large amounts of polyunsaturated oils was stimulated out of interest in the health benefits of lowering the amount of saturated fat, such as from butter and lard, in our diets. The main sources of polyunsaturated oil, which are sunflower, safflower, canola, and corn oils, are substantially unbalanced in terms of the *omega-6* to *omega-3* ratios. Those oils contain mostly *omega-6* fatty acids.

Animal studies have shown that in comparison with the *omega-3* fatty acids in fish oil, *omega-6* fatty acids can have a detrimental effect on insulin sensitivity[28]. Increased intake of *omega-3* fatty acids such as eicosapentaenoic acid (EPA) and docosahexaenoic acid (DHA), particularly when it is used to partially replace *omega-6* fatty acids, increases sensitivity to insulin and decreases serum triglycerides[29]. High intake of *omega-3* fatty acids decreases the risk of Metabolic Syndrome[30]. Although *omega-3* fatty acids tend to increase serum HDL levels, they may also increase levels of LDL. This has been a point of concern for some physicians. Nonetheless, increased intake of *omega-3* fatty acids is strongly associate with reductions in cardiovascular disease[31]. The benefits to health are substantial.

Fish oils that are rich in the *omega-3* fatty acids EPA and DHA can be important in reducing risk of Metabolic Syndrome[32]. The advantage of fish oil over flax seed oil, which is rich in *alpha*-linolenic acid, is simple to explain. Although the body can synthesize EPA and DHA from *alpha*-linolenic acid, an *omega-3* fatty acid found in some vegetable oils, it is not terribly efficient in doing so. It is better to consume EPA and DHA in fatty ocean fish or supplements than to rely on their production from *alpha*-linolenic acid.

Aside from their roles as energy storage molecules and building blocks of cell membranes, fatty acids are necessary to produce a number of critical chemical messengers in the body. The powerful, hormone-like prostaglandins, leukotrienes, thromboxanes and related substances are known collectively as eicosanoids. They are all produced from metabolites of the *omega-3* fatty acid alpha-linolenic acid and the *omega-6* fatty acid arachidonic acid.

Although both *omega-3* and *omega-6* fatty acids can be transformed into eicosanoids that mediate inflammatory processes in the body, the resulting eicosanoids can be different from each other. Though I hesitate to over-simplify, the resulting eicosanoids tend to antagonize each other's effects, with *omega-6* fatty acids supporting inflammation, and *omega-3* ameliorating it[33]. As I will discuss in Chapter 3, inflammation plays an important role in the progression of Metabolic Syndrome.

TRANS FATS

Even single ingestions of *trans* fatty acids decrease sensitivity to insulin and increase serum glucose levels[34]. *Trans* fatty acids worsen every aspect of the serum lipid profile. They increase LDL and decrease HDL. They raise serum triglycerides[35] and reduce the particle size of LDL[36]. *Trans* fatty acids also exacerbate oxidative stress. In mouse studies, high intake of *trans* fatty acids depletes body supplies of the antioxidant vitamin E, and increases levels of prostaglandins associated with inflammation and tissue damage[37].

Trans fats cause a variety of problems that further the progress of Metabolic Syndrome. Concern about *trans* fatty acids in the diet has increased to the point that Denmark has enacted laws to ban commercially produced *trans* fatty acids from the food supply. New York City has joined them in banning *trans* fat from city restaurants, and Canada is considering such legislation[38]. The effects of *trans* fatty acids can be strongly associated with Metabolic Syndrome and cardiovascular disease. Increased intake of *trans* fatty acids has been linked to increased risk for a first heart attack[39]. Every 5 g increase in daily intake of *trans* fatty acid increases the risk of coronary artery disease by 25 percent[40]. In view of the substantial increase in risk for heart attack that results from 5 g of *trans* fatty acid a day, it is worthwhile to consider what levels of *trans* fatty acid a common western diet may include. A report in the *New England Journal of Medicine* showed that in one fast-food restaurant in New York City, a meal that included French fries and a deep fried filet of chicken contained over 5.5 g of *trans* fatty acid. In another New York fast-food outlet, a similar meal contained over 10 g[41]. High levels of *trans* fatty acids are found in margarine, crackers, pancake mixes, corn chips, commercially produced pastries, and many other foods. Thankfully, it has been reported that *trans* fatty acids leave fatty tissues of the body fairly quickly after the supply of these nasty molecules has been cut off.

When I was an undergraduate in organic chemistry (quite a few years ago, I am afraid) I was interested in *trans* fatty acids in the diet. In particular, I was interested in whether or not frying with unsaturated oils in metal utensils might not cause formation of *trans* fatty acids through high temperature and the action of metals as catalysts. After maintaining oleic acid at frying temperatures for hours in the presence of copper, I was able to identifiy the production of small amounts of the *trans* analogue of oleic acid, elaidic acid. I never published those data. Nonetheless, I think it might be worthwhile to determine the degree to which we inadvertently add *trans* fatty acids to our diets through misuse of otherwise healthy cooking oils.

MICRONUTRIENTS

Protein, carbohydrate, fat, and fiber make up the major portion of our dietary intake, and they are known as macronutrients. However, the micronutrients in our diet, that is vitamins and minerals, are also important to maintain a healthy energy balance and avoid Metabolic Syndrome.

Inadequate intake of calcium has been associated with decreases in insulin sensitivity[43]. On the other hand, increases in calcium intake have been statistically associated with decreases in the likelihood of developing Metabolic Syndrome[42]. Calcium supplementation helps control hypertension[44] and lower serum cholesterol[45]. Part of the benefits of adequate dietary calcium come from the ability of calcium to ionically bind fatty acids in the gut and form "soap." This reduces some of the absorption of dietary fat. It also binds to bile salts and prevents their reabsorption, which is likely the reason that calcium can help control serum cholesterol. Postprandial lipemia is reduced by calcium as found in dairy products[46]. This may be partially due to the other substances found in milk. However, calcium phosphate supplementation has similar effects. It is worth noting that some forms of calcium, such as calcium carbonate, do not offer this benefit, likely due to differences in solubility. Vitamin D enhances the beneficial effects of calcium. However, as I will later explain, it also offers its own unique benefits.

Adequate levels of magnesium are also protective[47]. There is a complicated relationship between insulin and magnesium. Insulin enhances transport of magnesium into cells, whereas increased intracellular levels of magnesium enhance response to insulin. In many individuals with diabetes type II, magnesium levels are low. This may be due to the inefficiency of insulin in helping bring magnesium into the cells. This initiates a vicious cycle, as magnesium remaining in the serum is then more likely to be excreted by the kidneys. It has been reported in numerous studies, that increasing intake of magnesium with daily supplementation decreases insulin resistance in sufferers of diabtetes type II. Magnesium also decreases levels of inflammatory C-reactive protein and other risk factors for Metabolic Syndrome[48].

Zinc plays a role in the metabolism of carbohydrate and fat. It may also enhance the binding of insulin to its receptor. There have been reports that supplementation of zinc can improve insulin sensitivity[49]. The minerals chromium and vanadium have particular interest in regard to Metabolic Syndrome. However, I will discuss those two substances at length in Chapter 10, the Supplement Primer section of this book.

Calcium, magnesium, and zinc are beneficial for insulin activity. However, high, though not necessarily abnormal, serum levels of iron can contribute to insulin resistance and, in turn, Metabolic Syndrome[50]. There are studies showing a direct relationship between the amount of iron stored in the body and the number of features of Metabolic Syndrome observed in experimental subjects[51]. There are even suggestions that increased levels of serum ferritin, a protein that stores iron, may be yet another biochemical marker for Metabolic Syndrome[52]. The mechanism by which iron worsens Metabolic Syndrome is not clearly established. It is also unclear whether patients developing Metabolic Syndrome should be instructed to reduce their intake of iron. Some authors have suggested that at least some of the ability of the so-called Mediterranean diet to reverse Metabolic Syndrome is due to less reliance on red meat and, consequently, less intake of iron[53]. Fish and other seafood tend to replace meat in this style of eating. Sirloin steak has four times, and beef liver nearly nine times the amount of iron in an equal amount of salmon. Be aware, however, that many seeds and vegetables are also rich sources of iron. For example, ounce for ounce, pumpkin seeds have about twice the iron as beef liver!

Vitamins are essential for energy metabolism. This is particularly the case for the B-complex vitamins. However, there are other aspects of vitamin activity with more direct consequence to Metabolic Syndrome. A high serum level of homocysteine has been recognized as an independent risk factor for cardiovascular disease. Although the mechanism remains unclear, hyperhomocysteinemia has also been associated with insulin resistance[54]. In one study, treatment with folic acid and vitamin B12, which are known to decrease serum homocysteine, was found to reduce both homocysteine and insulin resistance[55]. Niacin also has ability to reverse some of the adverse effects of Metabolic Syndrome[56]. I will discuss niacin more thoroughly in the chapter on nutritional supplements and Metabolic Syndrome.

Oxidative damage is a major adverse consequence of the poorly controlled burning of glucose and fat that occurs in Metabolic Syndrome. Unfortunately, the oxidative damage that results from Metabolic Syndrome goes on to further exacerbate the underlying pathophysiology of the syndrome in an ever-increasing spiral of toxicity. Oxidative damage and the inflammation it generates increase insulin resistance[57]. The insulin resistance further drives Metabolic Syndrome, which only adds to the problem of oxidative damage. There is no compelling evidence that the primary antioxidant vitamins C, E and D can directly improve insulin resistance. Still, there is good evidence that adequate dietary or supplemental intake of antioxidants, such as vitamins C, E and D, can help reduce oxidative damage as well as slow down the progression of Metabolic Syndrome[58]. Vitamin D also has significant benefits in the treatment of psychiatric disorders, and I discuss it at length in Chapter 10.

SALT

It has long been known that too much salt can cause increases in blood pressure, which is a major component of Metabolic Syndrome. There are complex interactions among salt, insulin, hypertension and Metabolic Syndrome. Insulin increases sodium retention, and this effect is enhanced in people with Metabolic Syndrome[59]. This results in an enhanced ability of sodium to increase blood pressure in patients with Metabolic Syndrome. This phenomenon is known as salt sensitivity. Abdominal obesity, even without resistance to insulin, has also been associated with increases in salt sensitivity[60]. Several other factors common to abdominal obesity and Metabolic Syndrome, including increased serum levels of TNF[61] and leptin[62], are known to increase salt sensitivity.

Whereas Metabolic Syndrome can exacerbate the adverse effects of sodium, it remains to be determined if salt can affect the onset or progression of Metabolic Syndrome. Accounts in the medical literature vary. There is no typical response to either increase or restriction of sodium intake. The response may depend upon genetically determined sensitivity to salt, and whether or not an individual has already developed Metabolic Syndrome. High-salt diets can either increase[63] or reduce[64] resistance to insulin. One study found no effect of six days of high intake of salt on insulin resistance in healthy adults[65]. In another study, seven days of restriction of sodium intake to 3 g a day increased fasting insulin levels but had no effect on serum glucose or lipids[66].

There is evidence that too little salt can have adverse consequences on health, and possibly exacerbate components of Metabolic Syndrome. Very low salt intake can reduce blood pressure, and to compensate for this decrease, the adrenal glands can release more aldosterone. Levels of aldosterone can skyrocket by more than 300 percent[67]. Aldosterone, in turn, can induce insulin resistance and stimulate release of inflammatory adipocytokines from fat cells[68]. By and large, too little salt is rarely a problem.

In patients diagnosed as having Metabolic Syndrome, it is prudent to restict salt intake, with the degree of restriction matching the severity of the signs of the syndrome. Patients with four or five of the diagnostic criteria of

Metabolic Syndrome show a greater reduction of blood pressure in response to salt restriction than do those with only three criteria. In patients with only one, or none, of those diagnostic criteria, salt reduction has little if any effect on blood pressure[69]. In patients that have not shown signs of Metabolic Syndrome, simple moderation of salt intake is the most reasonable approach. Recent evidence shows that moderate reduction in salt intake, regardless of concern for Metabolic Syndrome, *per se*, can reduce cardiovascular risk by 25 percent[70].

EXERCISE

The increasingly sedentary lifestyle of our culture contributes to the current epidemic of obesity. However, the adverse effects of inactivity go beyond the mere failure to burn off extra calories. Physical exercise produces a variety of beneficial changes in physiology. In people who exercise regularly, sensitivity to insulin increases, fasting serum insulin decreases, and the ability to regulate glucose improves. Not unexpectedly, stopping physical training decreases insulin sensitivity[71]. This may be due to the fact that muscles of trained individuals have better blood supply and are able to extract glucose from blood more effectively. In such individuals, the liver also becomes more efficient in producing glycogen from the glucose it takes up. Changes in plasma enzymes that come with physical training further assist insulin sensitivity by enhancing clearance of triglycerides and fatty acids from the blood.

Exercise can also affect basal metabolic rate[72]. Basal metabolic rate is the amount of energy that must be used to simply stay alive in the absence of any physical activity. This is the energy needed to breathe, keep the heart beating, repair cells, keep the brain functioning, and so on. Being sedentary reduces, whereas being active increases, Basal Metabolic rate. Decreasing food intake can also decrease Basal Metabolic rate. This fact is very distressing for dieters, and may be a reason why decreasing food intake alone is often so ineffective in weight loss.

SMOKING AND DRINKING

Another lifestyle choice that increases the likelihood of developing Metabolic Syndrome is smoking. Smoking decreases insulin sensitivity. It also increases serum triglycerides and LDL, decreases serum HDL, and reduces LDL particle size[73]. Small particles of LDL are more likely to contribute to atherosclerosis. Serum levels of fibrinogen, which enhances blood clotting, are increased in smokers, as are levels of C-reactive protein and homocysteine[74]. All of those changes are associated with Metabolic Syndrome and increased risk of cardiovascular disease. Although people who quit smoking often gain weight, they still show improvements in serum cholesterol levels and insulin sensitivity[75].

Most studies have found that moderate amounts of alcohol on a regular basis can help reduce the risk for Metabolic Syndrome. In a recent study in the United States, the Third National Health and Nutrition Examination Survey, it was found that people who had 1 to 19 drinks a month had a 50 percent reduction in the likelihood of developing Metabolic Syndrome[76]. Among the benefits of moderate intake of alcohol were lower fasting insulin levels, increases in HDL, lower serum triglycerides, and smaller waist circumferences, which would suggest a decrease in visceral obesity. Drinking more than moderate amounts of alcohol begins to reduce those benefits. Red wine in particular has been touted as having unique health benefits. I will discuss red wine later in Chapter 10 in the section devoted to resveratrol. Resveratrol is thought to be one of the beneficial components of red wine. Nonetheless, in the study noted above, both wine and beer gave significant protection. Curiously, whiskey and other distilled alcoholic drinks did not provide significant health benefits.

OBESITY

Abdominal obesity is generally considered a major component of Metabolic Syndrome. However, the relationship between obesity and Metabolic Syndrome is a complex one. Obesity predisposes people to Metabolic Syndrome, and there is considerable overlap of the signs and symptoms in people that present these disorders. Nonetheless, obesity and Metabolic Syndrome are not identical. Obesity is defined as having an abnormally high proportion of fat. There are methods to determine how much fat a person has in their body. The most common office method of deciding whether or not a person is obese, aside from simply *looking* at them, is to determine the relationship between their height and weight. This relationship between weight and height is calculated as Body Mass Index (BMI)[77]. A normal BMI falls between 19 and 24.9. A BMI between 25 and 29.9 is overweight, whereas a BMI over 30 is considered obese. A BMI over 40 is described as severe obesity.

There are several problems with using BMI. First, the BMI makes no distinction between a muscle-bound weight lifter and a sedentary, flabby overeater. They could conceivably have the same weight and height. This is where common sense plays a role. The clinician can perform more specific testing, if necessary, to determine the actual proportion of fat in a patient. Another problem is that BMI makes no distinctions about where fat is distributed in the body. Women have a specific pattern of fat deposition. They tend to have fat accumulation in the hips, thighs and buttocks. It is the male pattern of abdominal fat deposition that is the most likely to cause health problems, including Metabolic Syndrome.

The fat cells in the abdomen are physiologically distinct from those in the hips, thighs and buttocks. It is visceral fat that contributes to the physiological changes of Metabolic Syndrome by secreting the pro-inflammatory and insulin-antagonistic adipocytokines. For these reasons, if a man and woman have the same elevated BMI, it is the man who would be most likely to show signs of Metabolic Syndrome. Of course, women do deposit visceral fat, especially under conditions of stress. They are not immune from developing Metabolic Syndrome.

Several other easily obtained measurements can provide information that BMI alone does not. One such measurement is simply the circumference around the waist at the level of the navel. The International Diabetes Federation considers waist circumferences of 94 cm (or 36.5 inches) in men and 80 cm (or 31.2 inches) in women to be indicative of Metabolic Syndrome. By and large, the relationship between BMI and waist circumference is close enough for BMI to be useful[78].

Perhaps the most significant fact to convey, is that obesity is only a risk factor for Metabolic Syndrome. It does not guarantee Metabolic Syndrome is present. In a recent study in the United States, it was found that 40 percent of adults with a BMI over 30 did not meet criteria for diagnosis of Metabolic Syndrome. Conversely, having a normal or even low BMI does not insure against Metabolic Syndrome. In a study from South Korea, a country in which less than 4 percent of adults are obese, roughly 15 percent of adults met criteria for Metabolic Syndrome[79].

Dr. Reaven has commented on the problem of trying to determine the significance of obesity in an individual who might be suffering Metabolic Syndrome. He has noted that while obesity does not entail insulin resistance, if an obese patient is insulin resistant, their response to insulin will almost certainly improve if they lose weight. Insulin resistance can be measured by rather laborious means in a metabolic laboratory. However, Reaven has suggested that insulin resistance in obese patients can be estimated by considering the ratio of triglycerides to HDL-cholesterol. Ratios above 3 suggest insulin resistance, and such patients should be strongly encouraged to lose weight and reduce intake of carbohydrates[80].

When obese, nondiabetic patients lose weight, there can be significant restoration of the body's sensitivity to insulin, as well as subsequent decreases in serum triglycerides[81]. Other improvements from weight loss in obese individuals include reductions in serum glucose, free fatty acid, and fasting serum insulin levels[82]. Levels of C-reactive protein also tend to decrease, which suggests decreases in the level of inflammatory activity following weight loss[83]. Overall, the relationship between obesity and Metabolic Syndrome would lead the wise individual to avoid obesity altogether.

REFERENCES

1. Bogardus, E. et al., Relationship between degree of obesity and in vivo insulin action in man. *Am. J. Physiol.* 1985; 248:286–291.
2. Lilloja, S. et al., In vivo insulin action is a familial characteristic in non-diabetic Pima Indians. 1987; 36:1329–1335.
3. Storgaard, H. et al., Genetic and nongenetic determinants of skeletal muscle glucose transporter 4 messenger ribonucleic acid levels and insulin action in twins. *J. Clin. Endocrinol. Metab.* 2006; 91:702–708.
4. Neel, J.V., Diabetes mellitus: a "thrifty" genotype rendered detrimental by "progress"? *Am. J. Hum. Genet.* 1962; 14:353–362.
5. Coleman, D.L., Obesity genes: beneficial effects in heterozygous mice. *Science* 1979; 663–665.
6. Stern, M.P. and Mitchell, B.D., Genetics of insulin resistance. In *Contemporary Endocrinology: Insulin Resistance*, G. Reaven and A. Laws, eds. Humana Press Inc. Totowa, NJ, 1999, pp. 3–18.
7. Song, Q., Wang, S.S., and Zafari, A.M., Genetics of the Metabolic Syndrome. *Hosp. Physician* 2006; 42:51–61.
8. Pooley, E.C. et al., A 5-HT2C receptor promoter polymorphism (HTR2C-759C/T) is associated with obesity in women, and with resistance to weight loss in heterozygotes. *Am. J. Med. Genet. B Neuropsychiatr. Genet.* 2004; 126:124–127.

9. Chaki, S. and Okuyama, S., Involvement of melanocortin-4 receptor in anxiety and depression. *Peptides* 2005; 26:1952–1964.

10. Lerer, B. et al., Variability of 5-HT2C receptor cys23ser polymorphism among European populations and vulnerability to affective disorder. *Mol. Psychiatry* 2001; 6:579–85.

11. Jenkins, D.J. et al., Glycemic index of foods: a physiological basis for carbohydrate exchange. *Am. J. Clin. Nutr.* 1981; 34:362–366.

12. McKeown, N.M. et al., Carbohydrate nutrition, insulin resistance, and the prevalence of the metabolic syndrome in the Framingham Offspring Cohort. *Diabetes Care* 2004; 27:538–546.

13. Hudgins, L., Effects of high-carbohydrate feeding on triglyceride and saturated fatty acid synthesis. *Proc. Soc. Exp. Biol. Med.* 2000; 225:178–183.

14. Vessby, B., Dietary fat, fatty acid composition in plasma and the metabolic syndrome. *Curr. Opin. Lipidol.* 2003; 14:15–19.

15. Parks, E.J. and Hellerstein, M.K., Carbohydrate-induced hypertriacylglycerolemia: Historical perspective and biological mechanisms. *Am. J. Clin. Nutr.* 2000; 71:412–423.

16. Crapo, P.A. et al., The effects of oral fructose, sucrose, and glucose in subjects with reactive hypoglycemia. *Diabetes Care* 1982; 5:512–517.

17. Okano, G. et al., Effect of pre-exercise fructose ingestion on endurance performance in fed men. *Med. Sci. Sports Exerc.* 1988; 20:105–109.

18. Elliott, S.S. et al., Fructose, weight gain, and the insulin resistance syndrome. *Am. J. Clin. Nutr.* 2002; 76:911–922.

19. Jeppesen, J. et al., Postprandial triglyceride and retinyl ester responses to oral fat: effects of fructose. *Am. J. Clin. Nutr.* 1995; 61:787–791.

20. Teff, K. et al., Dietary fructose reduces circulating insulin and leptin, attenuates postprandial suppression of ghrelin, and increases triglycerides in women. *J. Clin. Endocrinol. Metab.* 2004; 89:2963–2972.

21. Jenkins, D.J.A. et al., Dietary fiber, lente carbohydrates and the insulin-resistant diseases. *Br. J. Nutr.* 2000; 83 (Suppl. 1):S157–S163.

22. Bonanome, A. and Grundy, S.M., Effect of dietary stearic acid onplasma cholesterol and lipoprotein levels. *N. Engl. J. Med.* 1988; 318:1244–1248.

23. Tremblay, A.J. et al., Associations between the fatty acid content of triglyceride, visceral adipose tissue accumulation, and components of the insulin resistance syndrome. *Metabolism* 2004; 53:310–317.

24. Doucet, E. et al., Dietary fat composition and human adiposity. *Eur. J. Clin. Nutr.* 1998; 52:2–6.

25. Riccardi, G., Giacco, R., and Rivellese, A.A., Dietary fat, insulin sensitivity and the metabolic syndrome. *Clin. Nutr.* 2004; 23:447–456.

26. Mensink, R.P. and Katan, M.B., Effect of dietary fatty acids on serum lipids and lipoproteins. A meta-analysis of 27 trials. *Arterioscler. Thromb.* 1992; 12:911–919.

27. Simopoulos, A.P., Evolutionary aspects of diet, the *omega-6/omega-3* ratio and genetic variation: nutritional implications for chronic diseases. *Biomed. Pharmacother.* 2006; 60:502–507.

28. Jucker, B.M. et al., Differential effects of safflower oil versus fish oil feeding on insulin stimulated glycogen synthesis, glycolysis and pyruvate dehydrogenase flux in skeletal muscle. *Diabetes* 1999; 48:134–140.

29. Haag, M. and Dippenaar, N.G., Dietary fats, fatty acids and insulin resistance: short review of a multifaceted connection. *Med. Sci. Monit.* 2005; 11:359–367.

30. Warensjo, E. et al., Factor analysis of fatty acids in serum lipids as a measure of dietary fat quality in relation to the metabolic syndrome in men. *Am. J. Clin. Nutr.* 2006; 84:442–448.

31. Balk, E.M. et al., Effects of *omega-3* fatty acids on serum markers of cardiovascular disease risk: a systematic review. *Atherosclerosis* 2006; 189:19–30.

32. Connor, W.E., DeFrancesco, C.A., and Connor, S.L., N-3 fatty acids from fish oil. Effects on plasma lipoproteins and hypertriglyceridemic patients. *Ann. N.Y. Acad. Sci.* 1993; 683:16–34.

33. Calder, P.C., N-3 polyunsaturated fatty acids and inflammation: from molecular biology to the clinic. *Lipids* 2003; 38:343–352.

34. Lefevre, M. et al., Comparison of the acute response to meals enriched with cis- or *trans*-fatty acids on glucose and lipids in overweight individuals with differing FABP2 genotypes. *Metabolism* 2005; 54:1652–1658.

35. Mensink, R.P. et al., Effects of dietary fatty acids and carbohydrates on the ratio of serum total to HDL cholesterol and on serum lipids and apolipoproteins: a meta-analysis of 60 controlled trials. *Am. J. Clin. Nutr.* 2003; 77:1146–1155.

36. Mauger, J.F. et al., Effect of different forms of dietary hydrogenated fats on LDL particle size. *Am. J. Clin. Nutr.* 2003; 78:370–375.

37. Cassagno, N. et al., Low amounts of *trans* 18:1 fatty acids elevate plasma triacylglycerols but not cholesterol and alter the cellular defence to oxidative stress in mice. *Br. J. Nutr.* 2005; 94:346–352.

38. Mozaffarian, D. et al., *Trans* fatty acids and cardiovascular disease. *N. Engl. J. Med.* 2006; 354:1601–1613.

39. Clifton, P.M., Keogh, J.B., and Noakes, M. *Trans* fatty acids in adipose tissue and the food supply are associated with myocardial infarction. *J. Nutr.* 2004; 134:874–879.

40. Oomen, C.M. et al., Association between *trans* fatty acid intake and 10-year risk of coronary heart disease in the Zutphen Elderly Study: a prospective population-based study. *Lancet* 2001; 357:746–751.

41. Stender, S. et al., High levels of industrially produced *trans* fat in popular fast foods. *N. Engl. J. Med.* 2006; 354:1650–1652.

42. Liu, S. et al., Dietary calcium, vitamin D, and the prevalence of metabolic syndrome in middle-aged and older U.S. women. *Diabetes Care* 2005; 28:2926–2932.

43. Hagstrom, E. et al., Serum calcium is independently associated with insulin sensitivity measured with euglycemic-hyperinsulinaemic clamp in a community-based cohort. *Diabetologia* 2007; 50:317–324.

44. Van Mierlo, L.A. et al., Blood pressure response to calcium supplementation: a meta-analysis of randomized controlled trials. *J. Hum. Hypertens.* 2006; 20:571–580.

45. Ditscheid, B., Keller, S., and Jahreis, G., Cholesterol metabolism is effected by calcium phosphate supplementation in humans. *J. Nutr.* 2005; 135:1678–1682.

46. Lorenzen, J.K. et al., Effect of dairy calcium or supplemental calcium intake on postprandial fat metabolism, appetite, and subsequent energy intake. *Am. J. Clin. Nutr.* 2007; 85:678–687.

47. Barbagallo, M. et al., Role of magnesium in insulin action, diabetes and cardio-metabolic syndrome X. *Mol. Aspects Med.* 2003; 24:39–52.

48. Song, Y. et al., Magnesium intake, C-reactive protein, and the prevalence of metabolic syndrome in middle-aged and older U.S. women. *Diabetes Care* 2005; 28:1438–1444.

49. Marreiro, D.N. et al., Role of zinc in insulin resistance. *Arq Bras Endocrinol. Metabol.* 2004; 48:234–239.

50. Bozzini, C. et al., Prevalence of body iron excess in the metabolic syndrome. *Diabetes Care* 2005; 28:2061–2063.

51. Bozzini, C. et al., Prevalence of body iron excess in the metabolic syndrome. *Diabetes Care* 2005; 28:2061–2063.

52. Zelber-Sagi, S. et al., NAFLD and hyperinsulinemia are major determinants of serum ferritin levels. *J. Hepatol.* 2007; 46:700–707.

53. Panagiotakos, D.B. and Polychronopoulos, E., The role of Mediterranean diet in the epidemiology of metabolic syndrome; converting epidemiology to clinical practice. *Lipids Health Dis.* 2005; 4:7–14.

54. Bjorck, J. et al., Associations between serum insulin and homocysteine in a Swedish population-a potential link between the metabolic syndrome and hyperhomocysteinemia: the Skaraborg project. *Metabolism* 2006; 55:1007–1013.

55. Setola, E. et al., Insulin resistance and endothelial function are improved after folate and vitamin B12 therapy in patients with metabolic syndrome: relationship between homocysteine levels and hyperinsulinemia. *Eur. J. Endocrinol.* 2004; 151:483–489.

56. Meyers, C.D. and Kashyap, M.L., Management of the metabolic syndrome – nicotinic acid. *Endocrinol. Metab. Clin. North Am.* 2004; 33:557–575.

57. Evans, J.L., Maddux, B.A., and Goldfine, I.D. The molecular basis for oxidative stress-induced insulin resistance. *Antioxid. Redox Signal.* 2005; 7:1040–1052.

58. Ford, E.S. et al., The metabolic syndrome and antioxidant concentrations. Findings from the third National Health and Nutrition Examination Survey. *Diabetes* 2003; 52:2346–2352.

59. Skott, P. et al., Effect of insulin on renal sodium handling in hyperinsulinemic typeII (non-insulin-dependent) diabetic patients with peripheral insulin resistance. *Diabetologia* 1991; 34:275–281.

60. Strazzullo, P. et al., Altered renal sodium handling in men with abdominal adiposity: a link to hypertension. *J. Hypertens.* 2001; 19:2157–2164.

61. DiPetrillo, K. et al., Tumor necrosis factor induces sodium retention in diabetic rats through sequential effects on distal tubule cells. *Kidney Int.* 2004; 65:1676–1683.

62. Haynes, W.G., Role of leptin in obesity-related hypertension. *Exp. Physiol.* 2005; 90:683–688.

63. Donovan, D.S. et al., Effect of sodium intake on insulin sensitivity. *Am. J. Physiol.* 1993; 264:E730–E734.

64. Melander, O., Groop, L., and Hulthén, U.L. Effect of salt on insulin sensitivity differs according to gender and degree of salt sensitivity. *Hypertension* 2000; 35:827–831.

65. Foo, M. et al., Effect of salt-loading on blood pressure, insulin sensitivity and limb blood flow in normal subjects. *Clin. Sci. (Lond.)* 1998; 95:157–164.

66. Lind, L. et al., Metabolic cardiovascular risk factors and sodium sensitivity in hypertensive subjects *Am. J. Hypertens.* 1992; 5:502–505.

67. Jürgens, G. and Graudal, N.A., Effects of low sodium diet versus high sodium diet on blood pressure, renin, aldosterone, catecholamines, cholesterols, and triglyceride. *Cochrane Database Syst. Rev.* 2004; 1:CD004022.

68. Kraus, D. et al., Aldosterone inhibits uncoupling protein-1, induces insulin resistance, and stimulates proinflammatory adipokines in adipocytes. *Horm. Metab. Res.* 2005; 37:455–459.

69. Hoffman, I.S. and Cubeddu, L.X., Increased blood pressure reactivity to dietary salt in patients with the metablic syndrome. *J. Hum. Hypertens.* 2007 (Epublished ahead of print).

70. Cook, N.R. et al., Long term effects of dietary sodium reduction on cardiovascular disease outcomes: observational follow-up of the trials of hypertension prevention (TOHP). *BMJ* 2007 (Epublished).

71. Dela, F. et al., Physical activity and insulin resistance in man. In *Contemporary Endocrinology: Insulin Resistance*, G. Reaven and A. Laws, eds. Humana Press Inc., Totowa, NJ, 1999, pp. 97–120.

72. Molé, P.A., Impact of energy intake and exercise on resting metabolic rate. *Sports Med.* 1990; 10:72–87.

73. Eliasson, B. and Smith, U., Insulin resistance in smokers and other long-term users of nicotine. In *Contemporary Endocrinology: Insulin Resistance*, G. Reaven and A. Laws, eds. Humana Press Inc., Totowa, NJ, 1999, pp. 121–136.

74. Bazzano, L.A. et al., Relationship between cigarette smoking and novel risk factors for cardiovascular disease in the United States. *Ann. Intern. Med.* 2003; 138:891–897.

75. Eliasson, B. et al., Smoking cessation improves insulin sensitivity in healthy middle-aged men. *Eur. J. Clin. Invest.* 1997; 27:450–456.

76. Freiberg, M.S. et al., Alcohol consumption and the prevalence of the Metabolic Syndrome in the US: a cross-sectional analysis of data from the Third National Health and Nutrition Examination Survey. *Diabetes Care* 2004; 27:2954–2959.

77. Garrow, J.S., Three limitations of body mass index. *Am. J. Clin. Nutr.* 1988; 47:553.

78. Zhu, S. et al., Percentage body fat ranges associated with metabolic syndrome risk: results based on the third National Health and Nutrition Examination Survey (1988–1994). *Am. J. Clin. Nutr.* 2003; 78:228–235.

79. Park, H.S. et al., The metabolic syndrome and associated lifestyle factors among South Korean adults. *Int. J. Epidemiol.* 2004; 33:328–336.

80. Reaven, G., The insulin resistance syndrome: definition and dietary approaches to treatment. *Annu. Rev. Nutr.* 2005; 25:391–406.

81. Olefsky, J.M. et al., Effects of weight reduction on obesity: studies of carbohydrate and lipid metabolism. *J. Clin. Invest.* 1974; 53:64–76.

82. McLaughlin, T. et al., Metabolic changes following sibutramine-assisted weight loss in obese individuals: role of plasma free fatty acids in the insulin resistance of obesity. *Metabolism* 2001; 50:819–824.

83. McLaughlin, T. et al., Differentiation between obesity and insulin resistance in the association with C-reactive protein. *Circulation* 2002; 106:2908–2912.

3
THE PATHOPHYSIOLOGY OF METABOLIC SYNDROME

INSULIN

The peptide hormone insulin has an important and unique role in animal physiology. Its structure and mechanism of action have remained virtually unchanged over hundreds of millions of years. Insulin-like signaling processes can be seen in animals as primitive as the lowly *Caenorhabditis elegans* worm. Even the insulin of fish, quite distant from us in evolution, can be effective in humans. Insulin has had a lot of time to become complicated.

Insulin is released from the beta cells of the pancreas after intake of food. A number of nutrients in the blood can stimulate its release. However, glucose is the most important one. Insulin's major role is to switch the body from a fasting to an absorptive state of metabolism. Insulin turns off the liver's release of glucose, which it produces from glycogen and gluconeogenisis, and prepares it to take up glucose and store it as glycogen. Insulin also stimulates the liver to synthesize triglycerides and package them in VLDL for transport to the adipocytes for storage as fat. It increases the synthesis of albumin and other proteins by the liver, which deprives the gluconeogenesis pathways of amino acid substrate. It has been estimated that as much as half of insulin's ability to reduce postprandial glucose levels is due to its action on the liver.

Aside from the liver, insulin's effects are most dramatic on muscle and fat cells. Insulin binds to receptors on myocytes and causes them to increase their uptake of glucose. This increase is largely the result of insulin stimulating the translocation of type IV glucose transporters (GLUT IV) to the cell membranes. There is also a shift in the metabolism of myocytes from the burning of fatty acids to the burning of glucose. Muscle, like liver tissue, can store glucose in the form of glycogen. Insulin stimulates the synthesis and storage of glycogen in myocytes. This is the basis of so-called "carb loading" that athletes perform prior to long and strenuous events. Amino acids, which could otherwise be turned into glucose by the liver, are taken up more readily by the insulin-stimulated myocytes and used to produce various proteins. Insulin also inhibits the enzymatic breakdown of protein in myocytes, which further reduces the likelihood of amino acids being used simply as a source of glucose.

In adipocytes, insulin also increases the uptake of glucose and the utilization of amino acids for protein synthesis. Most significantly, insulin increases the synthesis of triglycerides and decreases the breakdown of fat and the secretion of fatty acids into the blood. The resulting drop in serum fatty acid levels forces the myocytes to burn glucose instead of fatty acid, which is their preferred fuel. Although insulin may stimulate somewhat different chains of events in liver, muscle, and fat tissue, the overall effect is the same. Insulin enhances the uptake and burning of glucose, and the storage of calories for future use. The storage process has the extra benefit of depriving the body of energy sources other than glucose. In this way, insulin lowers postprandial serum glucose to normal levels and helps store calories for future use.

Like most chemical messengers in the body, insulin acts first by binding to its own unique receptors in the cell membrane. However, what is unusual about insulin is the extraordinarily complicated system of secondary signals it triggers within the target cells it stimulates. Dozens of different proteins are phosphorylated in cascade fashion after insulin binds to the cell[1]. These proteins scatter throughout the cell and mediate activities along a variety of biochemical pathways. Some effects of insulin, such as activation of enzymes for utilizing glucose as fuel are almost immediate. Other effects of the hormone, including some mediated by genomic mechanisms, may occur over hours or days. A consequence of this complexity is that there are many secondary pathways along which errors in processing can occur to diminish response to insulin.

INSULIN RESISTANCE AND COMPENSATORY HYPERINSULINEMIA

Insulin resistance, *per se*, is a fairly straightforward concept. It is simply the state in which abnormally large amounts of insulin are needed to maintain normal serum glucose levels. It should be emphasized that serum glucose can be perfectly normal in patients with insulin resistance. That is, insulin resistance does not entail glucose intolerance. It all depends upon whether or not the pancreas can keep up with the higher levels of insulin that are needed to correct serum glucose. When the pancreas can no longer put out sufficient insulin to lower glucose, then the patient crosses the threshold into diabetes type II. Lipotoxicity and glucotoxicity, which I will soon discuss, are largely responsible for damage to the pancreas that prevents this organ from keeping up with the body's demand for insulin.

The enormous complexity of insulin resistance lies in the fact that so many abnormalities of physiology can cause it. Defects in the insulin receptor that prevent adequate stimulation by insulin are the purest and most obvious forms of insulin resistance. However, genetic defects in the insulin receptor itself are rare, and many, such as leprechaunism[2], are fatal early in life. It is not so uncommon for antiinsulin receptor antibodies to be present and in competition with insulin for binding at the receptor. This results in insulin resistance, as higher insulin levels are required to successfully compete for the receptor and produce the desired effect on serum glucose. Defects in the insulin receptor signal proteins that mediate the intracellular effects of insulin receptor activation can cause a number of problems that manifest as insulin resistance. One example is the hindrance of the translocation of GLUT IV transporters to the outer membranes of target cells. The paucity of GLUT IV transporters in the cell membrane decreases the rate at which glucose can be taken up into cells, and thus increases insulin requirement. Chronic hypercortisolemia leads to persistent increases in serum glucose and, in turn, a demand for more insulin. Thus, hypercortisolemia produces its own unique form of insulin resistance. Any of dozens of pathophysiological states that interfere with the storage of calories, the control of gluconeogenesis by the liver, or the uptake and burning of glucose in the cells can manifest as insulin resistance.

In Metabolic Syndrome the body suffers from insensitivity to insulin. However, it also suffers from the compensatory hyperinsulinemia that arises as the pancreas pumps out the increasingly larger amounts of insulin required to overcome the body's resistance to the hormone. Because insulin stimulates the autonomic nervous system, the hyperinsulinemia of Metabolic Syndrome contributes to chronic hypertension. At high concentrations, insulin can mimic the effects of the similarly structured hormone, insulin-like growth factor. Persistent, uncontrolled stimulation of insulin-like growth factor receptors in Metabolic Syndrome can stimulate overgrowth of the media of arterial walls[3], which further contributes to hypertension and cardiac disease. Stimulation of insulin-like growth factor receptors by insulin in ovarian tissue may play a role in the progression of polycystic ovary disease, which is an illness characterized by insulin resistance and hyperinsulinemia.

An important fact to recognize is that insulin resistance is not necessarily homogenous throughout the body. Whereas insulin sensitivity may be compromised in certain target cells, the hormone may act in perfectly normal fashion in other cells. Even within a single cell, one secondary insulin signaling system may remain intact whereas another may be defective. Simply stated, when insulin resistance is present, some cellular processes may be understimulated by insulin, whereas others may be overstimulated.

THE RISE OF THE ADIPOCYTE

Historically, doctors have viewed adipocytes as the three-toed sloths of human cell biology. They have been characterized as dumb, inert warehouses of fat. However, over the last 10 years there has been an explosion of data showing that adipocytes are as active, complicated, and communicative as any other cells in the body. The classical, primary role of adipocytes is to store fat and dispense it when it is needed. However, fat is a potentially dangerous substance, and a more recently recognized and equally important function of adipocytes is to protect other cells in the body from developing lipotoxicity.

Adipocytes sense the presence and need for fat, and they coordinate the use and intake of fat by other cells in the body. In serving this role, the adipocytes must stay in close communication with other cells. They are exquisitely sensitive to chemical signals from other cells. They have in their cell membranes receptors not only for insulin, but also for ACTH, thyroid stimulating hormone, growth hormone, prolactin, oxytocin, vasopressin and other hormones[4]. At the same time, adipose tissue acts as an endocrine gland not unlike the thyroid, adrenals, testes, and ovaries. This endocrine-like activity is performed primarily by visceral adipocytes. The abdominal obesity that helps define Metabolic Syndrome is due to expansion and overfilling of visceral adipocytes[5], which can disturb their endocrine-like functions.

Visceral adipocytes secrete a variety of potent hormones and hormone-like substances that are referred to as adipocytokines. There is a rapidly growing list of adipocytokines that can affect other cells of the body. Among the most important of these are adiponectin, tumor necrosis factor-alpha, IL-6, and leptin. The list also includes resistin, visfatin, and other substances that have been discovered only in the last few years. Adipocytokines affect not only the metabolism of fat and carbohydrate throughout the body, but the brain and behavior as well. For example, adiponectin secreted into the blood by adipocytes can directly affect the brain to cause changes in appetite and body weight. Adiponectin also affects the release of corticotrophin releasing factor, the brain hormone that plays a major role in the intiation of the stress response and stimulation of cortisol secretion.

Tumor necrosis factor alpha (TNF-alpha) is produced by a number of different cell in the body. It had been thought to come primarily from macrophages. It acquired its no-nonsense name by playing a part in the remission of cancers that can occur with activation of the immune system during serious bacterial infections. TNF-alpha stimulates inflammatory responses, and it is thought to be involved in psoriasis, rheumatoid arthritis and septic shock. It has only recently been discovered that TNF-alpha is produced in adipocytes. TNF-alpha levels are increased in patients with Metabolic Syndrome, and it is suspected of contributing to insulin resistance[6]. It is likely that the increase of fat burden in visceral adipocytes is responsible for the increase in TNF-alpha in the Metabolic Syndrome.

Interleukin 6 (IL-6) is one of a number interleukins produced in adipocytes. High serum levels of IL-6 are found in patients with Metabolic Syndrome, and levels vary with the degree of insulin resistance[7]. IL-6 levels increase in inflammatory states, and it is likely that IL-6 contributes to inflammatory damage in Metabolic Syndrome. As I will later explain (in Chapter 10) serum levels of IL-6 and TNF-alpha are increased in patients with Major Depression. It is likely that these substances are involved in the interrelationships between Metabolic Syndrome, Major Depression and inflammatory states.

LEPTIN

In 1977, it was discovered that normal mice and a strain of obese mice (called *ob*, for obesity) differed by a mutation of a single gene[8]. It was later found that the defective gene in the *ob* mouse was active almost exclusively in fat cells[9]. The *ob* mice not only became obese, but also developed hyperglycemia and insulin resistance. When the protein product of the normal gene was injected into the *ob* mice, they ate less, lost weight, and regained normal blood glucose and insulin levels[10]. The protein was most effective when administered directly into the brain. In fact, when the ventromedial hypothalamus was destroyed, the effects of leptin on food intake were substantially reduced[11]. The protein product of the normal *ob* gene was named leptin, which is derived from the Greek word for thin. Leptin seemed a likely candidate to be the long-suspected chemical signal sent from adipose tissue to the brain to help control food intake and energy metabolism.

All cells of the body can take up fatty acids. However, fatty acids are potentially toxic. One of leptin's major roles is to reduce accumulation of fatty acids in cells that cannot handle them safely[12]. When there is extra fat in the blood, adipocytes store it away while protecting other cells from lipotoxicity by releasing more leptin. Leptin stimulates the oxidation rather than the storage of fat in such cells, and prevents them from making more fats. Leptin may also stimulate PPAR[13], which is the target of many oral hypoglycemic medications. Indeed, the actions of leptin are intricately intertwined with those of insulin. Without leptin, people tend to become insulin resistant.

Because leptin restored the *ob* mouse to normal weight, it held promise as a treatment for human obesity. It was disappointing to find that administering leptin did not help obese patients lose weight[14]. In fact, it was soon discovered that most people with obesity and Metabolic Syndrome have high serum levels of leptin. As with insulin, the body can become resistant to leptin[15].

The physiological mechanism by which the body becomes resistant to leptin remains unclear. Insertion of leptin-producing genes directly into the brains of rats causes an overabundance of leptin inside the brain itself. However, these rats do not stop eating. Rather, their brains stopped responding to leptin. Apparently, too much leptin causes compensatory leptin resistance[66]. Leptin may shut itself off by stimulating one of the many Suppressor of Cytokine Signaling (SOCS) molecules[17] in the hypothalamic target cells. This increasingly important family of cytokine-regulating factors is also thought to play a role in insulin resistance and Metabolic Syndrome.

Resistance to leptin's effects on the brain may also be the result of a decrease in the ability of this adipocytokine to enter the brain. In obese patients, the amount of leptin in the brain is low in comparison with serum levels[18]. Several of the pathological changes seen in Metabolic Syndrome may contribute to this problem. High serum triglyceride, a defining feature of Metabolic Syndrome, makes it more difficult for leptin to cross the blood-brain barrier[19]. The inflammatory marker, C-reactive protein, whose levels are increased in Metabolic Syndrome, binds to leptin, blocks its effects, and adds to leptin resistance[20]. Dietary factors are also likely to add to this problem. Leptin resistance is a complicated phenomenon and, like insulin resistance, its ill effects are astonishingly widespread. Hyperleptinemia is now seen as a major risk factor for cardiovascular disease, diabetes, and hypertension[21]. As I will later discuss, leptin may also have effects on mood and sexual behavior.

ADIPOCYTES, INSULIN, AND METABOLIC SYNDROME

Insulin has a number of important effects on adipocytes. As it does in most other cells of the body, insulin stimulates the uptake of glucose in adipocytes by causing the translocation of GLUT IV transporters to the outer surface of their cell membranes. However, adipocytes do not consume a great deal of glucose. The increase in uptake of glucose by adipocytes accounts for only about 10 percent of insulin's hypoglycemic effect. This small percentage is trivial in comparison to the effects of insulin on glucose uptake by muscle cells.

The most important effects of insulin on adipocytes involve changes in the uptake, synthesis and storage of fat. Human adipocytes can synthesize fatty acids from glucose, and this process is facilitated by insulin[22]. They can also take up free fatty acids carried in the blood bound to albumin. However, most of the fatty acids that are stored in adipocytes come either from the diet, by way of the chylomicrons produced by the intestines, or from the liver in the VLDL that it synthesizes and releases into the blood. For the adipocyte to obtain fatty acids from the triglycerides of chylomicrons and VLDL, it must break them free from those large proteins with the enzyme lipoprotein lipase (LPL)[23]. LPL in human adipose tissue is enhanced by insulin[24]. This effect of insulin not only facilitates storage of fatty acids, but also furthers its own goals of lowering serum glucose by forcing muscle cells to use glucose instead of their preferred fuel of fatty acid. Inside the adipocyte, insulin enhances the synthesis of triglycerides from the fatty acids, while inhibiting the breakdown of triglycerides by the hormone-sensitive lipase (HSL) enzyme that resides in these cells[25]. HSL is used by adipocytes to break down triglyceride so that it can be excreted into the blood as free fatty acid to be used by the body as fuel between meals.

Adipocytes can become insulin resistant[26]. However, molecular engineering that selectively increases insulin resistance in adipocytes has little, if any immediate effect on insulin resistance in muscle and other tissues[27]. It is not that adipocytes are unimportant in maintaining insulin sensitivity. In patients with lipodystrophy, a condition in which adipocytes die off or simply fail to develop, insulin resistance is severe[28]. The most important role of adipocytes in preventing Metabolic Syndrome is the secretion of adipocytokines that helps other cells in the body manage fat and maintain sensitivity to insulin. Although insulin resistance in adipocytes can play a role in the pathology, it is primarily being overstuffed with fat that disturbs their ability to release helpful adipocytokines and inhibit release of potentially damaging ones. Overstuffed adipocytes produce less adiponectin, and decreases in adiponectin lead to insulin resistance. These adipocytes also overproduce leptin, which tends to precipitate leptin resistance. Adipocytes filled with too much fat also pour out inflammatory cytokines, such as IL-6 and TNF-alpha.

Genes that enhance fat deposition in adipocytes, and thus predispose to overstuffing adipocytes, are associated with insulin resistance in humans[29]. Many of these genes have been suspected of being among Neel's "Thrifty Genes". Genetic factors also determine the size adipocytes can reach and, ostensibly, the amount of fat thay can store before they are "overstuffed"[30]. Although insulin resistance in adipocytes may not be the fundamental cause of systemic insulin resistance and Metabolic Syndrome, the decrease in the activity of LPL and the disinhibition of HSL in insulin resistant adipocytes only make matters worse.

Inflammatory factors further contribute to the aberrant behavior of visceral adipocytes in Metabolic Syndrome. As visceral obesity increases, macrophages migrate into the adipose tissue. This may be due to the production of "monocyte chemoattractant protein-1" by adipocytes[31]. Adipocytes and macrophages secrete a variety of substances that alter the behavior and activity of each other. When these cell types are cultured together *in vitro*, the adipocytes secrete less adiponectin and the macrophages secrete more TNF-alpha[32]. The types of fatty acids present in the adipose tissue also affect the behavior of macrophages. Saturated fatty acids in particular increase their release of TNF-alpha and other inflammatory cytokines[33]. TNF-alpha then further increases insulin resistance in adipocytes[34] and disinhibits the activity of HSL in those cells[35]. This enhancement of HSL increases the breakdown of fat and the secretion of fatty acids into the blood, which increases lipemia and furthers the progression of Metabolic Syndrome in other cells of the body. This interaction is only one of many ways by which the Metabolic Syndrome accelerates out of control.

The level of "fight or flight" activation of the sympathetic nervous system also plays a role in the pathophysiological processes of adipocytes that aggrevate insulin resistance and Metabolic Syndrome. Adipocytes have adrenoceptors, and they are important targets of the adrenaline released under conditions of stress or persistent sympathetic nervous system activation. Activation of adrenoceptors on adipocytes increases the breakdown and release of fatty acids, even in the presence of insulin[36]. It is yet another form of insulin resistance. The subtype of adrenergic receptor that mediates the effects of sympathetic stimulation of adipocytes is thought to be the beta-3 adrenergic receptor. Certain genetic variations in this subtype of adrenergic receptor can predispose patients to obesity and Metabolic Syndrome.[37]

MUSCLE

The major problem in skeletal muscle cells that contributes to Metabolic Syndrome is improper handling of fat. Too much fat is taken up and, once inside the cell, it is not efficiently routed into oxidation pathways. There are several reasons that too much fat can be taken up into muscle cells. In the simplest cases, there is too much fat in the blood. This can be the result of eating too much fat, particularly saturated fats. However, excess calories obtained from carbohydrate are transformed into the saturated fatty acid, palmitate. In many individuals, high carbohydrate diets can cause marked increases in serum fatty acids and triglycerides. Indeed, that was Ahren's discovery in 1961 that led Jerry Reaven to investigate the phenomenon of insulin resistance. Those individuals are at risk for increases in fatty acid uptake into skeletal muscle.

In some people, there may be a genetic predisposition to take up fat too readily. Again, this harkens back to Neel's "Thrifty Gene" theory. One of the likely culprits is the CD36 fat-transporter protein. It has been found that certain subtypes of CD36 protein predispose patients to obesity and Diabetes II[38]. Adipocytokines can also effect the manner in which skeletal muscle cells take up fat. Leptin is usually depended upon to prevent the accumulation of fat in muscle cells. It does so, in part, by regulating the activity of CD36[39]. However, in states of leptin resistance this regulatory function of leptin is poorly performed. This can result in muscle cells taking up more fat than they can handle.

In Metabolic Syndrome, muscle cells also fail to properly shunt fat into oxidation pathways. Too much fatty acid, particularly palmitate, gets partitioned into so-called nonoxidative storage[40]. Adipocytokines released from visceral fat alter the way in which myocytes partition fatty acids for oxidation versus storage. Resistin and TNF-alpha decrease oxidation of fatty acids, whereas adiponectin and leptin enhance it[41]. Serum levels of restin and TNF-alpha increase in Metabolic Syndrome[42]. In contrast, serum adiponectin levels drop as the body's fat burden increases. Patients with Metabolic Syndrome tend to have low serum levels of adiponectin[43]. Leptin levels increase along with the amount of fat stored in adipocytes. However, as with insulin, the body can become resistant to its effects. Both hyperleptinemia and leptin resistance are common findings among those diagnosed with Metabolic Syndrome.

Accumulation of saturated fatty acids in skeletal muscle cells is associated with insulin resistance[44]. However, whereas increased uptake of palmitate into muscle cells increases insulin resistance, uptake of monounsaturated oleic acid does not[45]. This may be due in part to the fact that accumulation of saturated fatty acids leads to the production of toxic substances such as ceramides in these cells. Ceramides interfere with intracellular insulin signaling proteins[46]. Ceramides also stimulate the production of inflammatory mediators, such as Nuclear Factor kappa Beta (NFk-B), TNF-alpha, and IL-6. These substances decrease synthesis of GLUT IV glucose transporters and, consequently, insulin-stimulated glucose uptake[47]. Animal studies have shown that increases in *omega-3* fatty acid intake can decrease the amount of fatty acid that accumulates in muscle cells and contributes to insulin resistance[48].

A substantial portion of insulin's ability to lower serum glucose arises from its ability to increase uptake of glucose into skeletal muscle to be used as fuel. Insulin resistance in skeletal muscle adds a substantial burden to maintaining glycemic control. Thus, insulin resistance in muscle leads to hyperinsulinemia and, in many individuals, hyperglycemia. Skeletal muscle is also prone to damage from hyperglycemia through the usual problems of oxidative damage, protein glycosylation and cross-linking. Generally, this type of damage occurs late in the Metabolic Syndrome when the pancreas is no longer able to produce enough insulin to maintain glycemic control[49].

THE LIVER

The liver is the metabolic workhorse of the body. In the postprandial state, it stores away extra glucose molecules by stringing them together as glycogen or transforming them into fatty acid through the process of lipogenesis. The liver also converts the free fatty acids it takes up from the blood into triglycerides and releases them back into the blood to be distributed to adipocytes for storage. During the fasting state, the liver supplies the body with glucose that it produces either by synthesizing it *de novo* through the process of gluconeogenesis or by retrieving it from its glycogen stores through the process of glycolysis.

Insulin acts on the liver to the same ends that it acts on muscle and fat cells. That is, its goal is to reduce postprandial serum glucose levels back to normal while helping to store extra calories for future consumption. Insulin accomplishes this goal in the liver by stimulating glycogenesis and lipogenesis while inhibiting gluconeogenesis and glycolysis.

In hepatocytes, resistance can occur at any number of points along the complicated insulin signaling system. The variety and complexity of the biochemical functions performed by the liver make it the ideal place for the manifestation of the "hit or miss" quality of insulin resistance. The enzyme systems that mediate the generation of glucose through the pathways of gluconeogenesis and glycolysis become resistant to the inhibitory effect of insulin. On the other hand, enzymes in the liver that generate fatty acids out of carbohydrate by lipogenesis continue to be stimulated by insulin, even in animals with clear signs of systemic insulin resistance[50].

However, the effects of insulin on the synthesis and release of triglygeride-rich VLDL are complex and, at times, seemingly contradictory. For example, while insulin stimulates lipogenesis, there is compelling evidence that it may inhibit synthesis of the protein components of VLDL. This effect of insulin would decrease release of triglyceride-laden VLDL into the blood in normal individuals[51]. After prolonged exposure to insulin, such as would be found in the hyperinsulinemia of Metabolic Syndrome or from the stimulation of insulin release due to snacking on potato chips all day, the ability of the hormone to decrease VLDL synthesis decreases[52]. Thus, hypertriglyceridemia may be more an effect of insulin resistance through the phenomenon of tachyphylaxis of hepatic insulin receptors, rather than an isolated persistence of insulin sensitivity. In some respects, this is a twist on Reaven's initial suspicion that insulin stimulation *per se* increases triglyceride-rich VLDL secretion and causes hypertriglyceridemia.

THE LIVER, FAT, AND INSULIN RESISTANCE

There is something about the accumulation of excess fat in the liver that diminishes its response to insulin. In insulin-dependent diabetic patients, about 60 percent of the variance in insulin sensitivity is attributable to increased levels of fat in the liver[53]. Treatments that enhance the oxidation or block the synthesis of fatty acids in the liver also enhance the ability of insulin to inhibit gluconeogenesis and maintain serum glucose levels[54]. Indeed, when insulin resistant patients with diabetes type II lose weight, they also regain hepatic sensitivity to insulin. This return of insulin sensitivity correlates with the amount of decrease in intrahepatic fat[55].

Adipocytes store fat and monitor its metabolism throughout the body. However, in resistant adipocytes, insulin is less effective in inhibiting the enzyme HSL. The disinhibition of HSL, which breaks down triglyceride into component fatty acids, results in an increase of secretion of fatty acids into the blood. Insulin resistance in visceral adipocytes is particularly significant, as they empty their fat directly into the portal system that feeds the liver. The liver is then forced to deal with the extra free fatty acid that it must either oxidize or convert to triglyceride-laden VLDL. Extra carbohydrates in the diet are also turned into fat in the liver by lipogenesis. In animal models of hepatic insulin resistance, both high-fat and high-carbohydrate diets have been used to cause fatty liver and insulin resistance.

It is not entirely clear how the accumulation of fat in the form of triglycerides leads to insulin resistance. Fatty acids themselves can interfere with the binding of insulin to its receptor in the liver[56]. However, this does not explain the partial insulin resistance that is exhibited by the liver. That is, the persistence of insulin-stimulated lipogenesis with resistance to inhibition of gluconeogenesis and glycolysis. Perhaps the accumulation of fat affects only specific pathways in the intracellular insulin signaling system.

Some of the insulin resistance caused by fat in the liver may be due to the inflammatory effects fat tends to generate. When expression of the inflammatory factor NFk-B is inhibited in the liver, production of gluconeogenetic enzymes is curtailed and insulin sensitivity is increased[57]. The farnesoid X receptor (FXR) might also mediate some of the effects of fat on insulin activity in the liver. FXR is seen as a molecular link between fat and glucose metabolism in the liver[58]. Activation of FXR increases glycogenesis and decreases gluconeogenesis in the liver[59]. Animals lacking in FXR exhibit hypertriglyceridemia[60].

Whereas the accumulation of fat can contribute to insulin resistance, the type of fat being stored may also be significant. Polyunsaturated fatty acids in the diet decrease the activity of sterol regulatory element-binding protein-1, which is a molecular trigger for lipogenesis in the liver[61]. They also stimulate PPAR, which facilitates the burning of fatty acid in the liver. Polyunsaturated, but not saturated, fatty acids also have unique ability to modulate activity of the FXR receptor and thereby increase insulin sensitivity and reduce hyperlipidemia[62].

NON-ALCOHOLIC FATTY LIVER DISEASE

The increase in fat in the blood passing through the liver, along with the insulin-driven increase in lipogenesis, results in increases in both the synthesis of triglycerides and the release of triglyceride-rich VLDL into the blood. This results in deposition of fat and, sooner or later, lipotoxicity to occur in tissues throughout the body. In many individuals, there can also be pathological accumulation of triglyceride in the liver itself. This condition is known as nonalcoholic fatty liver disease (NAFLD)[63]. NAFLD is estimated to occur in up to 24 percent of people in the general population, and it has come to be seen as a common and expected hepatic manifestation of Metabolic Syndrome.

As fat begins to accumulate in the liver, a host of secondary problems emerge. Insulin resistance is only one such problem. Hepatocytes swelling with fat begin to impede the flow of blood through the sinusoids and impair function. Damage also results from poorly controlled oxidation of fat and inflammation. In 10 percent of patients with NAFLD, the condition progresses to nonalcoholic steatohepatitis. In a few unfortunate individuals, there can be still further progression to cirrhosis and hepatocellular malignancy.

HORMONAL INTERACTIONS IN THE LIVER

The effects of insulin on the liver are altered in Metabolic Syndrome. However, other hormones, including leptin, adiponectin and cortisol, can affect the way in which the liver processes calories and responds to insulin.

Patients with lipodystrophy have few if any adipocytes, and their serum leptin levels are quite low. These patients almost invariably have insulin resistance and fatty livers[64]. Leptin facilitates the regulation of gluconeogenesis by insulin and reduces the accumulation of triglycerides in hepatic tissue. On the other hand, elevated leptin concentrations have been found to block insulin's ability to reduce the process of gluconeogenesis in the liver[65]. Whether this is the result of too much activation by leptin or too little activation secondary to leptin resistance is not known. It is known that the interactions of leptin and insulin in the liver are extremely complex. Some effects of insulin on hepatocytes are enhanced by leptin, whereas others are inhibited[66].

Adiponectin is produced by adipocytes, but levels decrease as adipocytes become overfilled with fat. Adiponectin helps the liver handle fat and respond to insulin. Low serum levels of adiponectin are associated with higher severity of NAFLD[67]. Cortisol stimulates gluconeogenesis in the liver, and would thus be expected to increase insulin resistance and hasten the development of the NAFLD. However, hypercortisolemia has also been found to be an independent risk factor for NAFLD[68]. Interestingly, when hepatocytes are chronically exposed to high levels of both insulin and the glucocorticoid dexamethasone, their release of VLDL increases dramatically. Chronic stress and Metabolic Syndrome is a dangerous combination.

PEROXISOME PROLIFERATION ACTIVATING RECEPTORS

The peroxisome proliferation activating receptor (PPAR)[69] is a nuclear receptor in the steroid receptor family. It exists in three major subtypes and is active in many different cell types throughout the body. Its most important effects occur in muscle, adipose and liver tissue. PPAR is best known from the fact that the "glitazone" family of diabetic medication acts to stimulate its activity. This class of medication works over weeks or months to reverse insulin resistance and help prevent the progression of diabetes type II. The glitazones and other activators of PPARs may also hold promise for fighting cardiovascular disease and inflammatory conditions[70].

The primary role of PPAR is to regulate the oxidation of fat within cells. PPAR increases fatty acid uptake into cells, enhances burning of fat by beta-oxidation, and activates uncoupling proteins[71]. The latter proteins allow for the burning of fat for heat rather than chemical energy, and this facilitates the disposal of fat in states of overabundance. One of the causes of insulin resistance is the buildup of fatty acids in muscle cells. Because PPAR helps to rid these cells of extra fatty acid, it helps to maintain insulin sensitivity[72]. Not surprisingly, PPAR and leptin work together in performing this duty, and they appear to have reciprocal interactions. It is worth noting that one of the ways that fructose causes insulin resistance is by interfering with the ability of leptin to stimulate PPAR in the liver[73].

The natural ligands for PPAR are fatty acids and eicosanoids. In many respects, PPAR is a fatty acid sensor that brings information directly to the genome so that cells are prepared to deal with changes in the availability of fatty acid fuel. There are, however, marked differences in the ways various fatty acids affect PPAR. The *omega-3* fatty acid, EPA, in fish oil stimulates the synthesis of mRNA for production of the important gamma form of PPAR in human adipose tissue[74]. This does not occur after exposure to *omega-6* fatty acids, or even the *omega-3* fatty acid, DHA. Moreover, although high concentrations of most fatty acids increase the activation of PPARs, the saturated fatty acid palmitate and monounsaturated oleate are far less effective than are polyunsaturated acids[75]. In keeping with its other deleterious effects on the body, *trans* fatty acids inhibit the activity of PPARs[76].

INFLAMMATION

Inflammation is an extremely complicated series of chemical and cellular reactions the body performs in its attempt to defend itself against a variety of insults. Inflammation is a nonspecific form of body defense that likely predates the more sophisticated and precise actions of the immune system, such as the production of antibodies. Although many of the aspects of the inflammatory response are relatively primitive and nonspecific, evolution has completely and seamlessly integrated them into the most complex immune system activities.

There are common acute inflammatory responses, such as the release of histamine in response to tissue irritation. Moreover, almost any type of damage to tissue releases the fatty acid arachidonic acid from the cell membrane. Arachidonic acid is quickly converted by enzymes into inflammatory prostaglandins and leukotriens. These, in turn, attract cells to clot blood and fight infection, all the while increasing pain and swelling. Aspirin and drugs like ibuprofen block the conversion of arachidonic acid into those inflammatory eicosanoids.

When initial attempts fail to stop irritation and infection, the body hunkers down for a longer struggle with a different set of substances and defending cells. Primary among them are the immune cells called macrophages from the Greek for "big eaters". The macrophages take up residence in the damaged area, and release a stew of powerful chemicals including tumor necrosis factor, platelet activating factor, interleukins, eicosanoids, and many others.

As is often the case in physiology, a fine balance exists between the benefits the inflammatory response can offer and the damage it has potential to do. In Metabolic Syndrome, the inflammatory system is out of balance. A number of substances are released into the blood and tissues that set off both acute and chronic inflammatory responses. Some of what trips the alarms of inflammation are products of oxidative damage that arise out the poorly controlled processes of energy metabolism. Inflammation is also generated by proteins that become damaged by too much glucose in body fluids. Overburdened fat cells are stimulated to release inflammatory adipocytokines, many of which are the same inflammatory substances released from macrophages.

The compensatory hyperinsulinemia of Metabolic Syndrome also stimulates the inflammatory system. Intravenous infusion of insulin, with glucose administered along with it to maintain a normal serum glucose level, increases levels of the inflammatory cytokines interleukin-1beta, interleukin-6 and TNF-alpha in both the blood and cerebrospinal fluid[77]. The fact that cerebrospinal fluid shows signs of inflammatory process in the compensatory hyperinsulinemia of Metabolic Syndrome indicates that the brain is affected. I will later discuss how this may be one of the mechanisms by which Metabolic Syndrome contributes to psychiatric illness and dementia.

Dietary indiscretions can aggrevate the subtle, but persistent inflammatory processes of Metabolic Syndrome. Saturated fatty acids, including palmitate and the somewhat less villainous stearate, stimulate inflammatory effects in the endothelium and other tissues. This occurs by activation of the nuclear inflammation trigger NFk-B[78]. Oleic and other unsaturated fatty acids, such as linoleic acids, can ameliorate the stimulation of NFk-B by saturated fat. Along with adverse effects on serum lipids, *trans* fatty acids increase signs of inflammation, such as increase in serum levels of TNF, IL-6, and C-reactive protein[79].

OXIDATIVE STRESS

The word stress is used in many contexts and for many different reasons. Thus, the term "oxidative stress" may sound a bit vague. However, through the work of scientists such as Dr. Michael Brownlee, who, like Dr. Reaven, has been a recipient of the Banting Prize, oxidative stress has now been quite clearly defined in its nature, origin and impact upon the body. The oxidative stress with which Dr. Brownlee has concerned himself, is the increase in abnormal oxidation that occurs in states of hyperglycemia. Hyperglycemia quite often occurs in patients with diabetes. However, it may also occur in Metabolic Syndrome and help set the stage for further progression to diabetes type II.

The underlying physiology of the oxidation of glucose in the mitochondria is complicated. However, it is something that has been studied for many years, and most of the process is well known by physiologists and biochemists. Dr. Brownlee explains[80] in a simple but effective manner that when too much glucose is available, too much energy is unleashed, and the mitochondria can lose control of some of the dangerous side products of the glucose oxidation process. The most dangerous side product of this process is the oxygen free radical. It is an oxygen atom that has been given an extra electron, and it cannot wait to unload it on any unsuspecting molecule that happens to be nearby.

Usually, a protective enzyme called superoxide dismutase is around to change the oxygen radical into the less reactive substance, hydrogen peroxide. However, when there is an overabundance of glucose, the capacity of the system is overwhelmed. Aside from causing damage directly to cell structures, this oxidative stress sets into motion what Brownlee describes as "the four major pathways of hyperglycemic damage."

The mechanisms by which the four different pathways damage cells are ferociously complex. If the reader is of scientific bent, he or she may wish to read Dr. Brownlee's brilliant Banting Lecture, which sums up the work nicely without delving too deeply into any specific biochemical mechanisms. I will touch upon these pathways and their significance in the briefest way. The first mechanism, stimulation of the polyol pathway, results in the loss of the major natural antioxidant in cells, glutathione. This, in turn, makes the cells even more susceptible to the increasing oxidative stress. The liver is particularly dependent on the protective effects of glutathione. When conditions such as NAFLD already handicap the liver's ability to produce glutathione, further compromise of its availability by oxidative damage is doubly dangerous. The second mechanism is the covalent bonding of glucose with cell proteins and amino acids to produce what are called advanced glycation end-products (AGEs). When cells sense the presence of AGEs inside, they release inflammatory cytokines that initiate their usual chain reactions of damage.

A third pathway is the activation of the enzyme protein kinase C. The effects of protein kinase C activation are best summed up by Dr. Brownlee who says, "The things that are good for normal function are decreased and the things that are bad are increased." For the most part, the bad things include inflammation, hypertension, and further insulin resistance. The fourth pathway involves spillover of glucose into what is called the hexosamine pathway. Although this is a normal biochemical mechanism in the body, overactivation of the system results in substances that still further increase insulin resistance and damage the endothelium of arteries.

The damage from oxidative stress occurs in the cells that line blood vessels, and this, in turn, contributes to cardiovascular disease. The brain can also be ravaged by oxidative stress, which contributes to dementia and other degenerative diseases. The damage from hyperglycemia-induced oxidative stress also occurs in the beta cells of the pancreas, and when the damage is severe enough, the beta cells stop producing insulin. When the pancreas cannot produce enough insulin to maintain serum glucose levels, the patient officially suffers from diabetes type II.

The solution to the oxidative stress problem secondary to hyperglycemia is to reduce the amount of glucose available for burning in the mitochondria, and to provide as much antioxidant protection as possible to minimize the damaging effects of oxygen free radicals. Large fluctuations in serum glucose, such as occur when we snack throughout the day on high glycemic index foods, increase the risk of oxidative damage in the body. These fluctuations may be even more dangerous than sustained high levels of serum glucose[81]. Apparently, free radicals are produced in higher amounts by the mitochondria when large increases in serum glucose occur intermittently during the day. This oxidative stress further causes vascular damage, injury to pancreatic beta cells, and other damage in the body due to inflammation and so-called advanced glycation end products. This only further prompts a warning to avoid high glycemic index foods, and to snack on foods that tend to be high in protein and healthy fats.

LIPOTOXICITY

The oxidative damage described by Brownlee is the basis of the glucotoxicity that gradually whittles away at the integrity of the cardiovascular system, brain, pancreas, and other tissues of the body. However, in Metabolic Syndrome there is also a substantial amount of damage done by lipotoxicity. The primary reason that fat becomes so toxic in Metabolic Syndrome is that it ends up accumulating in tissues that are not prepared to handle it. This deposition of fat in the wrong places is due to leptin resistance, a lack of adiponectin, too much saturated fat in the diet, and the growing inability of insulin to inhibit the activity of HSL in adipocytes while continuing to stimulate lipogenesis in the liver[82].

Lipotoxicity is due to both oxidative and nonoxidative processes. In conditions of oxidative stress, tissues are exposed to relatively high levels of reactive free radicals. Lipids are particularly vulnerable to oxidation by these reactive species, and the result is lipid peroxidation. Lipid peroxidation stimulates inflammation, and may also participate in the pathogenesis of atherosclerosis. The oxidation of LDL, which plays a major role in the progression of atherosclerosis, is a phenomenon of lipid peroxidation.

The lipids that are most vulnerable to peroxidation are the *omega-6* fatty acids[83]. For reasons that are not entirely clear, *omega-3* fatty acids are less likely to be peroxidized than the *omega-6* fats. This has led to concern about the safety of increasing one's intake of foods rich in polyunsaturated fat. However, limiting polyunsaturated fats to no more than 10 percent of total fat intake; balancing the ratio of *omega-3* to *omega-6* in one's fat consumption; and maintaining an adequate supply of antioxidants, such as vitamins E and C, greatly minimizes this risk of peroxidation. Even when such care is not taken, the benefits of the polyunsaturated fatty acids, particularly as found in fish oil, may outweigh any risks.

There is also evidence that obesity itself may predispose to oxidative stress and lipid peroxidation independently of hyperglycemia[84]. This lipid peroxidation occurs in adipose tissue where fat accumulates. It may be due to interaction with the enzyme NADPH oxidase that resides in adipocytes. This peroxidation in adipose cells may occur early in Metabolic Syndrome, and it has been suspected that it plays a role in the dysregulated production and release of adipocytokines that causes the syndrome to progress. Contributions of lipid peroxidation and glucotoxic oxidative damage to the progress of Metabolic Syndrome may explain the fact that the syndrome is associated with low serum concentrations of the antioxidants vitamins A, C and E; carotenoids; and retinyl esters[85].

Some lipotoxicty in Metabolic Syndrome is created through nonoxidative processes. It occurs when tissues are overburdened with fat because of not enough fat being partitioned into oxidation pathways. The most important nonoxidative lipotoxic substances are the ceramides, which are produced through enzymatic action on palmitate and serine. Ceramides trigger a cascade of inflammatory signals, apparently beginning with NFk-B but including bad actors such as TNF-alpha[86]. They also stimulate production of highly reactive, peroxynitrites that, in some cells, act as trigger for apoptosis. Apotosis is sometimes described as the pathway of cellular suicide, in which chemical processes are initiated that eventually result in cell death. The way to minimize lipotoxicity due to ceramides is to reduce the fat burden, particularly that of saturated fat, and maintain sensitivity to insulin and leptin. It is worth noting that the "*cer*" in *cer*amide is the same as that in the word *cer*ebrum. There are high concentrations of ceramides in the normal brain. Defects in ceramide metabolism may play a part in the progression of Alzheimer's Disease[87]. However it is not known if the activity of ceramides in the brain is altered in Metabolic Syndrome.

ASYMMETRICAL DIMETHYLARGININE

A common feature of Metabolic Syndrome is impaired function of the arterial lining, or endothelium. Dysfunctional endothelium contributes to a variety of problems including hypertension, atherosclerosis, thrombosis, inflammation, myocardial infarction, and stroke. One abnormality of the endothelium in Metabolic Syndrome is a decrease in the synthesis and activity of nitric oxide (NO). Nitric oxide is generated in the endothelium and acts to relax arterial smooth muscle, increase blood flow, and decrease blood pressure. Nitroglycerin and other nitrate-based medications for hypertension and anginal pain act by stimulating NO synthesis in the endothelium.

Asymmetrical dimethylarginine (ADMA) and Symmetrical dimethylarginine (SDMA), are natural substances in the body that are generated during the normal cellular processes of proteolysis. Whereas SDMA is a relatively benign substance, ADMA acts as an NO synthesis inhibitor. Serum levels of ADMA are increased in a number of illnesses, including renal disease, hepatic illnesses, cardiovascular disease, and diabetes. It is coming to be seen as a risk factor for hypertension, atherosclerosis, and cardiac death[88].

Serum ADMA is also elevated in Metabolic Syndrome[89], although the reasons why are not entirely clear. This may occur as a function of increases in serum LDL cholesterol, as LDL upregulates the enzymes that produce ADMA during proteolysis[90]. Elevations may also come as the result of decreases in the enzymatic destruction of ADMA by dimethylarginine dimethylaminohydrolase (DDAH)[91]. Homocysteine, which is often elevated in Metabolic Syndrome, blocks the activity of DDAH.

As in Metabolic Syndrome, serum ADMA levels are also elevated in patients with Alzheimer's Dementia[92], Major Depression[93], and schizophrenia[94]. It is not known if ADMA plays any role in the initiation or progression of those psychiatric illnesses. Some of the neurodegeneration that occurs in Alzheimer's Dementia is believed to be due to oxidative damage from NO. Moreover, many antidepressants can, like ADMA, inhibit the enzyme nitric oxide synthase. Thus, there is no obvious connection between Alzheimer's Dementia, Major Depression, and decreases in NO activity that might be caused by increases in endogenous ADMA. On the other hand, some of the pathophysiology of schizophrenia is suspected to be due to decreases in the activity of NMDA receptors in the brain. The NMDA receptor, which uses NO as a second messenger in its intracellular signaling process, might be negatively affected by increased levels of ADMA in the body. This possibility will be discussed further in Chapter 4.

STRESS AND CORTISOL

Cortisol is one of the major steroid hormones released from the adrenal cortex. Cortisol is a glucocorticoid. This term is derived from the fact that one of the primary functions of cortisol is to maintain adequate serum glucose levels during the fasting state. It enhances serum glucose levels by enhancing the breakdown of protein and facilitating the transformation of the resulting amino acids into glucose by the process of gluconeogenesis. Cortisol also serves a permissive role in the body's maintenance of serum glucose levels. After cortisol has acted on the liver, the hormones glucagon and adrenaline are better able to stimulate the breakdown of glycogen and release of glucose from that organ.

At high concentrations, cortisol antagonizes some of the effects of insulin. It blocks the ability of insulin to enhance uptake of glucose into adipose and muscle cells. It also overrides insulin's ability to suppress the production and release of glucose by the liver. These effects of cortisol can manifest as insulin resistance. The main beneficiary of the hyperglycemic effects of cortisol is the brain. This is one reason why it is released under conditions of severe stress. Ostensibly, the brain needs the extra glucose to plot a course of action when being pursued by a lion, or to hatch a new strategy to keep the Internal Revenue Service at bay.

Cortisol can also act on adipose tissue as a means to marshall an adequate supply of energy during the fasting state. It enhances lipolytic activity in adipose tissue to increase serum free fatty acid levels and make the glycerol bound up in triglycerides available for conversion into glucose. Although cortisol helps to mobilize fat to be used during the fasting state, it also serves to stimulate the deposition of fat in visceral adipocytes[95]. The mechanism by which this occurs is not clear.

High serum cortisol, or hypercortisolemia, is not a defining feature of Metabolic Syndrome. However, hypercortisolemia, such as occurs in Cushing's disease, is generally accompanied by a constellation of changes very similar to those seen in Metabolic Syndrome. Patients with Cushing's disease often have hypertension, insulin resistance, hyperglycemia, hyperlipidemia, and visceral obesity[96].

Although adipose cells do not create steroid hormones from scratch, they contain the enzyme 11-beta-hydroxysteroid dehydrogenase type 1 (11-beta HSD-1) that converts the inactive metabolite cortisone back into the active parent steroid cortisol[97]. Because there is often substantial amounts of cortisone in the blood, adipose tissue can reintroduce significant amounts of the active stress hormone, cortisol, back into circulation. It has been shown that in people with visceral obesity, the ability of cortisol to increase insulin resistance and other aspects of the Metabolic Syndrome is enhanced[98]. This is likely due to the activity of 11-beta HSD-1. In fact, there is evidence that 11-beta HSD is overexpressed in patients with diabetes type II[99]. Overall, it appears that while cortisol exacerbates visceral obesity, visceral obesity can in turn contribute to hypercortisolemia. Both contribute to Metabolic Syndrome.

REFERENCES

1. Pessin, J.E. and Saltiel, A.R., Signaling pathways in insulin action: molecular targets of insulin resistance. *J. Clin. Invest.* 2000; 106:165–169.
2. Longo, N. et al., Genotype-phenotype correlation in inherited severe insulin resistance. *Hum. Mol. Gen.* 2002; 11:1465–1475.
3. Zavaroni, I. et al., Insulin resistance/compensatory hyperinsulinemia predict carotid intimal medial thickness in patients with essential hypertension. *Nutr. Metab. Cardiovasc. Dis.* 2006; 16:22–27.
4. Schaffler, A. et al., Hypothesis paper Brain talks with fat – evidence for a hypothalamic-pituitary-adipose axis? *Neuropeptides* 2005; 39:363–367.
5. Ritchie, S.A. and Connell, J.M., The link between abdominal obesity, metabolic syndrome and cardiovascular disease. *Nutr. Metab. Cardiovasc. Dis.* November 14, 2006 [Epub ahead of print].
6. Gomez-Fernandez, P. et al., [Biomarkers of vascular inflammation and subclinical atherosclerosis in the metabolic syndrome] *Med. Clin. (Barc.)* 2004; 123:361–363.
7. Helioavaara, M.K. et al., Plasma IL-6 concentration is inversely related to insulin sensitivity, and acute-phase proteins associate with glucose and lipid metabolism in healthy subjects. *Diabetes Obes. Metab.* 2005; 7:729–736.
8. Coleman, D.L., Obese and diabetes: two mutant genes causing diabetes-obesity syndromes in mice. *Diabetologia* 1978; 14:141–148.
9. Zhang, Y. et al., Positional cloning of the mouse obese gene and its human homologue. *Nature* 1994; 372:425–432.
10. Pelleymounter, M.A. et al., Effects of the obese gene product on body weight regulation in ob/ob mice. *Science* 1995; 269:540–543.
11. Satoh, N. et al., Pathophysiological significance of the obese gene product, leptin, in ventromedial hypothalamius (VMH)-lesioned rats: evidence for loss of its satiety effect in VMH-lesioned rats. *Endocrinology* 1997; 138:947–954.
12. Unger, R.H., Minireview: Weapons of lean body mass destruction – The role of ectopic lipids in the metabolic syndrome. *Endocrinology* 2003; 144:5159–5165.
13. Kakuma, T. et al., Role of leptin in peroxisome proliferator-activated receptor coactivator-1 expression. *Endocrinology* 2000; 141:4576–4582.
14. Veniant, M.M. and LeBel, C.P., Leptin: from animals to humans. *Curr. Pharm. Des.* 2003; 9:811–818.
15. Considine, R.V. et al., Serum immunoreactive-leptin concentrations in normal-weight and obese humans. *N. Engl. J. Med.* 1996; 334:292–295.
16. Zhang, Y. and Scarpace, P.J., The role of leptin in leptin resistance and obesity. *Physiol. Behav.* 2006; 88:249–256.
17. Ueki, K. et al., Central role of suppressors of cytokine signaling proteins in hepatic steatosis, insulin resistance, and the metabolic syndrome in the mouse. *Proc. Natl. Acad. Sci. USA* 2004; 101:10422–10427.
18. Caro, J.F. et al., Decreased cerebrospinal-fluid/serum leptin ratio in obesity: a possible mechanism for leptin resistance. *Lancet* 1996; 348:159–161.
19. Banks, W.A. et al., Triglycerides induce leptin resistance at the blood-brain barrier. *Diabetes* 2004; 53:1253–1260.
20. Chen, K. et al., Induction of leptin resistance through direct interaction of C-reactive protein with leptin. *Nat. Med.* 2006; 12:425–432.

21. Leyva, F. et al., Hyperleptinemia as a component of a metabolic syndrome of cardiovascular risk. *Arterioscler. Thromb. Vasc. Biol.* 1998; 18:928–933.
22. Moustaid, N., Jones, B.H., and Taylor, J.W., Insulin increases lipogenic enzyme activity in human adipocytes in primary culture. *J. Nutr.* 1996; 126:865–870.
23. Mead, J.R., Irvine, S.A., and Ramji, D., Lipoprotein lipase: structure, function, regulation and role in disease. *J. Mol. Med.* 2002; 80:753–769.
24. McTernan, P.G. et al., Insulin and rosiglitazone regulation of lipolysis and lipogenesis in human adipose tissue in vitro. *Diabetes* 2002; 51:1493–1498.
25. Large, V. et al., Metabolism of lipids in human white adipocyte. *Diabetes Metab.* 2004; 30:294–309.
26. Kashiwagi, A. et al., In vitro insulin resistance of human adipocytes isolated from subjects with noninsulin-dependent diabetes mellitus. *J. Clin. Invest.* 1983; 72:1246–1254.
27. Shi, H. et al., Overexpression of suppressor of cytokine signaling 3 in adipose tissue causes local but not systemic insulin resistance. *Diabetes* 2006; 55:699–707.
28. Sell, H., Dietze-Schroeder, D., and Eckel, J., The adipocyte-myocyte axis in insulin resistance. *Trends. Endocrinol. Metab.* 2006; 17:416–422.
29. Dubois, S.G. et al., Decreased expression of adipogenic genes in obese subjects with type 2 diabetes. *Obesity (Silver Spring)* 2006; 14:1543–1552.
30. Tittelbach, T.J. et al., Racial differences in adipocyte size and relationship to the metabolic syndrome in obese women. *Obes. Res.* 2004; 12:990–998.
31. Kamei, N. et al., Overexpression of monocyte chemoattractant protein-1 in adipose tissues causes macrophage recruitment and insulin resistance. *J. Biol. Chem.* 2006; 281:26602–26614.
32. Suganami, T., Nishida, J., and Ogawa, Y., A paracrine loop between adipocytes and macrophages aggravates inflammatory changes: role of free fatty acids and tumor necrosis factor alpha. *Arterioscler. Thromb. Vasc. Biol.* 2005; 25:2062–2068.
33. Todoric, J. et al., Adipose tissue inflammation induced by high-fat diet in obese diabetic mice is prevented by n-3 polyunsaturated fatty acids. *Diabetologia* 2006; 49:2109–2119.
34. Despres, J.P. and Marette, A., Obesity and insulin resistance. In *Insulin Resistance: The Metabolic Syndrome X*, Gerald Reaven Ami Laws, eds. Humana Press Inc., Totowa, NJ, 1999, pp. 51–81.
35. Langin, D. and Arner, P., Importance of TNFalpha and neutral lipases in human adipose tissue lipolysis. *Trends Endocrinol. Metab.* 2006; 17:314–320.
36. Jost, M.M. et al., The beta3-adrenergic agonist CL316,243 inhibits insulin signaling but not glucose uptake in primary human adipocytes. *Exp. Clin. Endocrinol. Diabet.* 2005; 113:418–422.
37. Malczewska-Malec, M. et al., An analysis of the link between polymorphisms of the beta2 and beta3 adrenergic receptor gene and metabolic parameters among Polish Caucasians with familial obesity. *Med. Sci. Monit.* 2003; 9:CR225–234.
38. Corpeleijn, E. et al., Direct association of a promoter polymorphism in the CD36/FAT fatty acid transporter gene with Type 2 diabetes mellitus and insulin resistance. *Diabet. Med.* 2006; 23:907–911.
39. Bonen, A. et al., Plasmalemmal fatty acid transport is regulated in heart and skeletal muscle by contraction, insulin and leptin, and in obesity and diabetes. *Acta Physiol. Scand.* 2003; 178:347–356.
40. Bell, J.A. et al., Dysregulation of muscle fatty acid metabolism in type 2 diabetes is independent of malonyl-CoA. *Diabetologia* 2006; 49:2144–2152.
41. Dyck, D.J., Heigenhauser, G.J., and Bruce, C.R., The role of adipokines as regulators of skeletal muscle fatty acid metabolism and insulin sensitivity. *Acta Physiol. (Oxf.)* 2006; 186:5–16.
42. Vettor, R. et al., Review article: adipocytokines and insulin resistance. *Aliment. Pharmacol. Ther.* 2005; 22 (Suppl. 2):3–10.
43. Tajtakova, M. et al., Adiponectin as a biomarker of clinical manifestation of metabolic syndrome. *Endocr. Regul.* 2006; 40:15–19.
44. Schrauwen-Hinderling, V.B. et al., Intramyocellular lipid content in human skeletal muscle. *Obesity (Silver Spring)* 2006; 14:357–367.
45. Bastie, C.C. et al., CD36 in myocytes channels fatty acids to a lipase-accessible triglyceride pool that is related to cell lipid and insulin responsiveness. *Diabetes* 2004; 53:2209–2216.

46. Lee, J.S. et al., Saturated, but not n-6 polyunsaturated, fatty acids induce insulin resistance: role of intramuscular accumulation of lipid metabolites. *J. Appl. Physiol.* 2006; 100:1467–1474.

47. Todd, M.K. et al., Thiazolidinediones enhance skeletal muscle triacylglycerol synthesis while protecting against fatty acid-induced inflammation and insulin resistance. *Am. J. Physiol. Endocrinol. Metab.* 2007; 292:E485–493.

48. Neschen, S. et al., Contrasting effects of fish oil and safflower oil on hepatic peroxisomal and tissue lipid content. *Am. J. Physiol. Endocrinol. Metab.* 2002; 282:E395–401.

49. Krebs, M. and Roden, M., Nutrient-induced insulin resistance in human skeletal muscle. *Curr. Med. Chem.* 2004; 11:901–908.

50. Shimomura, I., Bashmakov, Y., and Horton, J.D., Increased levels of nuclear SREBP-1c associated with fatty livers in two mouse models of diabetes mellitus. *J. Biol. Chem.* 1999; 274:30028–30032.

51. Malmstrom, R. et al., Metabolic basis of hypotriglyceridemic effects of insulin in normal men. *Arterioscler. Thromb. Vasc. Biol.* 1997; 17:1454–1464.

52. Bartlett, S.M. and Gibbons, G.F., Short- and longer term regulation of very-low-density lipoprotein secretion by insulin, dexamethasone and lipogenic substrates in cultured hepatocytes. *Biochem. J.* 1988; 249:37–43.

53. Ryysy, L. et al., Hepatic fat content and insulin action on free fatty acids and glucose metabolism rather than insulin absorption are associated with insulin requirements during insulin therapy in type 2 diabetic patients. *Diabetes* 2000; 49:749–758.

54. Savage, D.B. et al., Reversal of diet-induced hepatic steatosis and hepatic insulin resistance by antisense oligonucleotide inhibitors of acetyl-CoA carboxylases 1 and 2. *J. Clin. Invest.* 2006; 116:817–824.

55. Petersen, K.F. et al., Reversal of nonalcoholic hepatic steatosis, hepatic insulin resistance, and hyperglycemia by moderate weight reduction in patients with type 2 diabetes. *Diabetes* 2005; 54:603–608.

56. Navegantes, L.C. et al., Regulation and counterregulation of lipolysis in vivo: different roles of sympathetic activation and insulin. *J. Clin. Endocrinol. Metab.* 2003; 88:5515–5520.

57. Tamura, Y. et al., Ameliroation of glucose tolerance by hepatic inhibition of nuclear factor kappaB in db/db mice. *Diabetologia* 2007; 50:131–141.

58. Duran-Sandoval, D. et al., Potential regulatory role of the farnesoid X receptor in the metabolic syndrome. *Biochimie* 2005; 87:93–98.

59. Zhang, Y., Activation of the nuclear receptor FXR improves hyperglycemia and hyperlipidemia in diabetic mice. *Proc. Natl. Acad. Sci. USA* 2006; 103:1006–1011.

60. Sinal, C.J. et al., Targeted disruption of the nuclear receptor FXR/BAR impairs bile acid and lipid homeostasis. *Cell* 2000; 102:731–744.

61. Sekiya, M. et al., Polyunsaturated fatty acids ameliorate hepatic steatosis in obese mice by SREBP-1 suppression. *Hepatology* 2003; 38:1529–1539.

62. Zhao, A. et al., Polyunsaturated fatty acids are FXR ligands and differentially regulate expression of FXR targets. *DNA Cell Biol.* 2004; 23:519–526.

63. Neuschwander-Tetri, B.A., Fatty liver and nonalcoholic steatohepatitis. *Clin. Cornerstone* 2001; 3:47–57.

64. Javor, E.D. et al., Long-term efficacy of leptin replacement in patients with generalized lipodystrophy. *Diabetes* 2005; 54:1994–2002.

65. Petersen, K.F. et al., Leptin reverses insulin resistance and hepatic steatosis in patients with severe lypodystrophy. *J. Clin. Invest.* 2002; 109:1345–1350.

66. Cohen, B., Novick, D., and Rubinstein, M., Modulation of insulin activities by leptin. *Science* 1996; 15:1185–1188.

67. Targher, G. et al., Associations between plasma adiponectin concentrations and liver histology in patients with nonalcoholic fatty liver disease. *Clin. Endocrinol. (Oxf.)* 2006; 64:679–683.

68. Targher, G. et al., Relationship of non-alcoholic hepatic steatosis to cortisol secretion in diet-controlled Type 2 diabetic patients. *Diabet. Med.* 2005; 22:1146–1150.

69. Rangwala, S.M. and Lazar, M.A., Peroxisome proliferator-activated receptor gamma in diabetes and metabolism. *Trends Pharmacol. Sci.* 2004; 25:331–336.

70. Han, S.H., Quon, M.J., and Koh, K.K., Beneficial vascular and metabolic effects of peroxisome proliferator-activated receptor-alpha activators. *Hypertension* 2005; 46:1086–1092.

71. Luquet, S. et al., Roles of peroxisome proliferator-activated receptor delta (PPARdelta) in the control of fatty acid catabolism. A new target for the treatment of metabolic syndrome. *Biochimie* 2004; 86:833–837.

72. Hegarty, B.D. et al., The role of intramuscular lipid in insulin resistance. *Acta Physiol. Scand.* 2003; 178:373–383.

73. Roglans, N. et al., Impairment of hepatic Stat-3 activation and reduction of PPARalpha activity in fructose-fed rats. *Hepatology* 2007; 45:778–788.

74. Chambrier, C. et al., Eicosapentaenoic acid induces mRNA expression of peroxisome proliferator-activated receptor gamma. *Obes. Res.* 2002; 10:518–525.

75. Kliewer, S.A. et al., Fatty acids and eicosanoids regulate gene expression through direct interactions with peroxisome proliferator-activated receptors alpha and gamma. *Proc. Natl. Acad. Sci. U.S.A.* 1997; 94:4318–4323.

76. Saravanan, N. et al., Differential effects of dietary saturated and *trans*-fatty acids on expression of genes associated with insulin sensitivity in rat adipose tissue. *Eur. J. Endocrinol.* 2005; 153:159–165.

77. Fishel, M.A. et al., Hyperinsulinemia provokes synchronous increases in central inflammation and beta-amyloid in normal adults. *Arch. Neurol.* 2005; 62:1539–1544.

78. Staigler, K. et al., Saturated, but not unsaturated, fatty acids induce apoptosis of human coronary endothelial cells via nuclear factor-kappa Beta activation. *Diabetes* 2006; 55:3121–3126.

79. Mozaffarian, D. et al., Dietary intake of *trans* fatty acids and systemic inflammation in women. *Am. J. Clin. Nutr.* 2004; 79:606–612.

80. Brownlee, M., The pathobiology of diabetic complications. A unifying mechanism. *Diabetes* 2005; 54:1615–1625.

81. Monnier, L. et al., Activation of oxidative stress by acute glucose fluctuations compared with sustained chronic hyperglycemia in patients with type 2 diabetes. *JAMA* 2006; 295:1681–1687.

82. Unger, R.H. and Orci, L., Diseases of liporegulation: new perspective on obesity and related disorders. *FASEB* 2001; 15:312–321.

83. Eritsland, J., Safety considerations of polyunsaturated fatty acids. *Am. J. Clin. Nutr.* 2000; 71:197–201.

84. Furukawa, S. et al., Increased oxidative stress in obesity and its impact on metabolic syndrome. *J. Clin. Invest.* 2004; 114:1752–1761.

85. Ford, E.S. et al., The metabolic syndrome and antioxidant concentrations. Findings from the third national health and nutrition examination survey. *Diabetes* 2003; 52:2346–2352.

86. Summers, S.A., Ceramides in insulin resistance and lipotoxicity. *Prog. Lipid Res.* 2006; 45:42–72.

87. Mielke, M.M. and Lyketsos, C.G., Lipids and the pathogenesis of Alzheimer's Disease: is there a link? *Int. Rev. Psychiatry* 2006; 18:173–186.

88. Boger, R.H., The emerging role of asymmetric dimethylarginine as a novel cardiovascular risk factor. *Cardiovasc. Res.* 2003; 59:824–833.

89. Stuhlinger, M.C. et al., Relationships between insulin resistance and an endogenous nitric oxide synthase inhibitor. *JAMA* 2002; 287:1420–1426.

90. Boger, R.H. et al., LDL cholesterol up regulates synthesis of asymmetrical dimethylarginine in human endothelial cells. *Circ. Res.* 2000; 87:99–105.

91. Stuhlinger, M.C. et al., Homocysteine impairs the nitric oxide synthase pathway. Role of asymmetric dimethylarginine. *Circulation* 2001; 104:2569–2575.

92. Selley, M.L., Increased concentrations of homocysteine and asymmetric dimethylarginine and decreased concentrations of nitric oxide in the plasma of patients with Alzheimer's Disease. *Neurobiol. Aging* 2003; 24:903–907.

93. Selley, M.L., Increased (E)-4-hydroxy-2-noneal and asymmetric dimethylarginine concentrations and decreased nitric oxide concentrations in the plasma of patients with major depression. *J. Affect. Disord.* 2004; 80:249–256.

94. Das, I., Elevated endogenous nitric oxide synthase inhibitor in schizophrenic plasma may reflect abnormalities in brain nitric oxide production. *Neurosci. Lett.* 1996; 215:209–211.

95. Paulmyer-Lacroix, O. et al., Expression of the mRNA coding for 11beta-hydroxysteroid dehydrogenase type 1 in adipose tissue from obese patients: an in situ hybridization study. *J. Clin. Endocrinol. Metab.* 2002; 87:2701–2705.

96. Walker, B.R., Abnormal glucocorticoid activity in subjects with risk factors for cardiovascular disease. *Endocr. Res.* 1996; 22:701–708.

97. Mariniello, B. et al., Adipose tissue 11-beta hydroxysteroid dehydrogenase type 1 expression in obesity and Cushing's Syndrome. *Eur. J. Endocrinol.* 2006; 155:435–441.

98. Darmon, P. et al., Insulin resistance induced by hydrocortisone is increased in patients with abdominal obesity. *Am. J. Physiol. Endocrinol. Metab.* 2006; 291:995–1002.

99. Abdallah, B.M., Beck-Nielsen, H., and Gaster, M., Increased expression of 11beta-hydroxysteroid dehydrogenase type 1 in type 2 diabetic myotubes. *Eur. J. Clin. Invest.* 2005; 35:627–634.

4
METABOLIC SYNDROME AND PSYCHIATRIC ILLNESS

MAJOR DEPRESSION

It has long been known that a disproportionately large percentage of patients with diabetes also suffer Major Depression. The prevalence of Major Depression in diabetics, regardless of whether their diabetes is type I or type II, is roughly three times that seen in the general population[1]. However, the likelihood of depression is often increased in individuals dealing with serious and potentially disabling illnesses. Thus, the significance of the high prevalence of depression in diabetics, and whether it reflects some interaction between the two seemingly disparate conditions has not been entirely clear.

Over recent years it has become apparent that there is a relationship between Major Depression and Metabolic Syndrome, which is frequently the precursor of diabetes type II. Men and women with depression are more likely than those without depression to develop Metabolic Syndrome. People with depression often have the abdominal obesity, insulin resistance, hypertension, hyperlipidemia, and elevated fasting glucose levels that characterize the syndrome[2]. There is also a strong relationship between depression and insulin resistance[3]. Insulin resistance, the cardinal feature of Metabolic Syndrome, is four times more likely to occur in depressed individuals[4]. Although Major Depression is associated with Metabolic Syndrome, it is not clear how the two are related. Does depression cause Metabolic Syndrome, or do the biochemical changes of Metabolic Syndrome lead to depression?

Evidence suggests that depression can set the stage for later development of Metabolic Syndrome[5]. Women who complain of depression, anxiety, and anger are more likely than women without those psychological characteristics to develop Metabolic Syndrome in subsequent years. Path analyses have shown statistically that the progression from depression to poor health habits to Metabolic Syndrome is the most likely course of events when depression and Metabolic Syndrome coexist[6].

One explanation for how depression might lead to Metabolic Syndrome is that depression fosters unhealthy lifestyles[7]. Patients with depression often smoke. They are inactive and eat poorly. Although many patients with Major Depression have poor appetites, some sufferers crave sweets and indulge in "comfort foods" packed with high-glycemic-index-carbohydrates. This is particularly the case in so-called Atypical Depression[8]. Despite its name, Atypical Depression may actually be a common form of depression in women. Comorbidity of depression, obesity, and Metabolic Syndrome occurs more often in women than in men.

Overeating, smoking, drinking, inactivity, and carbohydrate craving are at least partially responsible for the tendency of people with depression to develop Metabolic Syndrome. However, even when these unhealthy habits are statistically removed from the equation, there is an unexpectedly high incidence of Metabolic Syndrome among sufferers of Major Depression[9]. It is possible that some of the same, underlying physiological abnormalities contribute to both Major Depression and Metabolic Syndrome.

INSULIN RESISTANCE AND MAJOR DEPRESSION

The insulin resistance of Metabolic Syndrome may play a role in Major Depression[10]. Although insulin dramatically affects the way muscle and fat cells take up and metabolize glucose, neuroscientists had come to believe that it had no significant effect on the brain's utilization of glucose. However, insulin has subtle, yet potentially important effects upon the way the brain uses glucose.

Insulin is actively transported into the brain, and insulin receptors are found in a variety of important areas of the brain[11]. The amount of insulin in the brain is affected by changes in serum insulin levels. Insulin levels in cerebrospinal fluid increase during acute episodes of hyperinsulinemia. In chronic hyperinsulinemia, such as occurs in Metabolic Syndrome, insulin levels in cerebrospinal fluid can decrease. This is likely due to down-regulation of the activity of the insulin transport system into the brain. When insulin is reduced to levels below those generally seen during the fasting state in humans, the brain uses less glucose[12]. It is the older areas of the brain, such as the cerebellum and brain stem, that have the highest density of insulin receptors. However, it is the cerebral cortex, where the higher functions of mind reside, that is most strongly affected by the depletion of insulin.

The insulin resistance that defines Metabolic Syndrome is most clearly seen in muscle, adipose tissue, and liver. Although changes in insulin level can clearly affect brain activity, there has been a question as to whether the insulin resistance seen in peripheral tissue also occurs in the brain tissue of patients with Metabolic Syndrome. Recent studies with human subjects have shown that this is the case. When glucose and insulin levels are controlled by clamping techniques, insulin resistant subjects show less cortical excitation than insulin sensitive subjects to the same serum levels of insulin. The degree of insulin's effect on cortical activity is in positive correlation with insulin sensitivity as measured by the ability of insulin to stimulate glucose uptake in peripheral tissues[13]. The ability of a specific level of serum insulin to enhance glucose metabolism in the cerebral cortex is also diminished in subjects with peripheral insulin resistance[14]. This resistance to insulin is most apparent in the prefrontal cortex and other areas of the brain involved in motivation and reward.

Perhaps the most compelling evidence of brain tissue becoming resistant to insulin comes from a study in which the response to insulin was evaluated *ex vivo* using slices of brain removed from hamsters made insulin resistant by feeding them a fructose-enriched diet. Direct administration of insulin into brain tissue was less effective in generating insulin-induced long term inhibition in sections from insulin resistant animals than in those from control animals[15]. Existing data do suggest that decreases in insulin activity in the brain, due either to insulin resistance in brain tissue or decreases in the ability of insulin to reach the brain, could play a significant role in the development of Major Depression.

Consistent with insulin resistance causing depression are findings that chromium, an essential mineral known to enhance the effects of insulin, can help relieve depression. Chromium has been used successfully for treatment of both depression[16] and dysthymia[17]. The latter is a mild but persistent form of depression. Interestingly, the type of depression that may best be helped by chromium is Atypical Depression. This is the form of depression more commonly seen in women, that is characterized by depressed mood, lack of motivation, low sex drive, carbohydrate craving, and sleeping too much. The strongest effects of chromium were in countering carbohydrate craving, increased appetite, and decreased sex drive[18].

Lithium is a medication that has long been used to treat Bipolar Affective Disorder. It is also used to augment the effects of antidepressants in treatment-resistant depression. Lithium mimics several of the effects of insulin in the brain. Both lithium and insulin inhibit the activity of the increasingly important enzyme, glycogen synthase kinase-3[19]. Moreover, both lithium and insulin stimulate the activities of the enzymes phosphatidylinositol-3-kinase and protein kinase B[20]. It is not known to what degree the benfits of lithium are due to mimicking the effects of insulin. Neither is it known if lithium has any special benefit for patients that suffer both depression and Metabolic Syndrome or diabetes type II. *Omega-3* fatty acids, which have been found to improve some symptoms of both Major Depression[21] and Metabolic Syndrome, have also been found to enhance the activity of phosphatidylinositol-3-kinase[22].

STRESS

Another common thread between Metabolic Syndrome and Major Depression are elevations in the level of the stress hormone, cortisol. Many people with severe depression have hypercortisolemia. Over prolonged periods of time, hypercortisolemia can cause a number of serious problems in the brain and throughout the body.

I was privileged to spend three years working with Dr. Bruce McEwen in his Laboratory of Neuroendocrinology at the Rockefeller University in New York. Dr. McEwen is one of the world's authorities on the effects of stress on the brain and body. It was McEwen that first discovered the presence of cortisol receptors in areas of the brain responsible for learning, memory, and emotion. This unexpected finding led scientists to understand that the increase in cortisol that occurs during stress does not simply affect the neurovegetative state. It affects activity throughout the brain, including its higher functions.

Cortisol is essential for adequate responses to the normal stresses of life[23]. However, when cortisol is elevated for prolonged periods of time, the body begins to suffer damage. Among the problems that can occur due to prolonged periods of stress are obesity, impairment of the immune system, atherosclerosis, and osteoporosis. There may even be atrophy of certain areas of the brain. Regrowth of neurons in the adult mammalian brain, a process known as neurogenesis, is disrupted by high levels of cortisol[24]. Stimulation of neurogenesis may play a role in the therapeutic response to antidepressants. Decreases in neurogenesis may predispose to depression and dementia. From 50–75 percent of patients with depression suffer from some degree of hypercortisolemia[25].

My own small contribution to the study of the effects of cortisol was the finding that hypercortisolemia leads to decreases in the number of serotonin type 1A (5-HT1A) receptors in the brain[26]. Decreases in serotonergic activity in the brain have been linked to depression, anxiety, and violence[27]. The 5-HT1A receptor mediates some of the antidepressant effects of serotonin[28]. It is likely that many antidepressant medications, particularly serotonin selective reuptake inhibitors (SSRIs), act by increasing activity at 5-HT1A receptors in the brain. A recent study using laboratory rats has shown that treatment with the antidepressants imipramine or citalopram reverses the adverse effects of corticosterone on 5-HT1A receptors in the brain[29].

In Cushing's disease, a disorder of the pituitary gland in which very large amounts of cortisol are secreted into the blood, patients can exhibit severe, and even psychotic forms of depression[30]. Psychotic depression can also result from treatment with high doses of prednisone and related steroids. I vividly recall consulting on a patient who was in the hospital for transplant surgery. He was receiving high doses of prednisone to prevent rejection. When I arrived he stared at me suspiciously and refused to answer my questions. Suddenly, he grabbed his shoes and began to methodically transfer them from one side of the bed to the other. With each placement of a shoe he proclaimed in alternating fashion, "Pretty shoes," "Ugly shoes," "Pretty shoes," "Ugly shoes." After a dozen such pronouncements, he exploded into bitter tears. He improved after starting antipsychotic medication and tapering down the prednisone. Interestingly, RU 486, a drug that blocks cortisol receptors in the brain, has recently been found to be useful in the treatment of severe, psychotic depression[31]. This is the same RU 486 that gained notoriety for possible use as an abortion drug. Cortisol and progesterone are steroids similar in structure, and RU 486 binds to both cortisol and progesterone receptors.

Hypercortisolemia also contributes to Metabolic Syndrome. There are many similarities between Metabolic Syndrome and the state of hypercortisolemia. In both there is impaired glucose uptake, increased fasting glucose levels, hyperlipidemia, deposition of visceral fat, and hypertension[32]. Although insulin sensitivity can be normal early in hypercortisolemia, the elevated cortisol levels eventually reduce insulin sensitivity[33]. Depressed patients with hypercortisolemia also tend to have increases in visceral fat that typifies Metabolic Syndrome[34]. This increase in visceral fat raises serum levels of inflammatory cytokines that exacerbate depression, insulin resistance, and Metabolic Syndrome. It also helps sustain high levels of serum cortisol through the activity of the enzyme 11-beta HSD-1 that converts inactive cortisone back into cortisol. 11-beta HSD-1 exists in high concentration in visceral adipocytes. Not surprisingly, there are data suggesting that mutations of the glucocorticoid receptor that mediates some of cortisol's effects can predispose people to both Major Depression[35] and Metabolic Syndrome[36].

INFLAMMATION

Another set of features common between Major Depression and Metabolic Syndrome is inflammation. People with Metabolic Syndrome have high levels of CRP, TNF-alpha, interleukin-6, and other substances that mediate the body's inflammatory response. Many individuals suffering major depression have high blood levels of the same inflammatory cytokines. Patients with Major Depression are more than twice as likely than those without depression to have elevated levels of CRP, IL-6, and TNF-alpha[37]. Inflammatory cytokines stimulate the hypothalamic-pituitary-adrenal (HPA) axis[38], with resulting increases in cortisol that exacerbate both depression and Metabolic Syndrome. Patients with depression, particularly when they have had a history of childhood emotional trauma, also have increased activity of NFk-B[39]. NFk-B is one of the major intracellular triggers of the inflammatory response. It mediates some of the damaging effects of oxidation and other pathological consequences of Metabolic Syndrome. Persistent activation of NFk-B in the brain can desensitize glucocorticoid receptors that mediate feedback inhibition of the HPA axis[40]. This leads to sustained states of hypercortisolemia due to disinhibition of the release of cortisol from the adrenal glands.

It is not yet clear whether depression causes an inflammatory response, or if it is the other way around. In all likelihood, each aggravates the other. In any case, a great deal of communication goes on between the brain and the cells of the immune system. Areas of the brain that are believed to be involved in Major Depression are quite sensitive to inflammatory cytokines. (For an excellent review you may wish to read the paper by Dr. Charles Raison *et al.*[41]) The brain not only responds to cytokines, but it also produces them to help initiate and control local immune and inflammatory responses. It is not the neurons themselves, but rather the attending glial cells that produce these cytokines in the brain[42]. Indeed, these cells can act, in many respects, like cells of the immune system.

Changes in the immune and inflammatory systems may be partially responsible for certain forms of Major Depression. Many of the symptoms of depression, including poor appetite, lack of energy, vague physical complaints, and lack of motivation overlap with those of a phenomenon called "Sickness Behavior". It has been discovered in recent years that some of the sick feelings in illness are the result of the body's own chemical defense mechanisms against disease. Although it sounds odd at first, it actually makes a great deal of sense. When ill, it is best to stay inside and rest. Because many illnesses come from eating contaminated food, it is also probably best not to eat anything for a while. Some cytokines cause fatigue, nausea, body aches, and a lack of desire to do things. This may be the body's way of making people do what is best for them when they are ill. Perhaps the best known example of a cytokine producing such effects is the severe depression many sufferers of hepatitis C or Multiple Sclerosis experience when treated with interferon[43]. However, other immune system cytokines, including IL-2 and TNF, can also cause fatigue and malaise. Thus, increases in certain inflammatory cytokines that occur in Metabolic Syndrome may worsen symptoms of depression by contributing to Sickness Behavior .

Dr. Michael Maes, one of the first physician-scientists to postulate a connection between depression and inflammation, has discovered that many antidepressants blunt the body's inflammatory response. Some antidepressants decrease the release of interferon while increasing the release of the anti-inflammatory cytokine, IL-10[44]. The antidepressant bupropion is able to reduce serum levels of TNF-alpha[45]. The relationship between inflammation and depression is further strengthened by the report that etanercept, a drug used to treat inflammatory diseases, can improve depression in those being treated[46]. Etanercept acts primarily as an antagonist at TNF-alpha receptors. Interestingly, there has been at least one report of a patient becoming manic after treatment with etanercept[47]. Antidepressants can trigger mania in susceptible individuals.

OBESITY

Obesity is a common component of Metabolic Syndrome. Although most people with obesity do not suffer depression, there is an increased risk for depression among the obese[48]. The risk of depression is much higher among obese women than obese men. In fact, men and women are quite different in this respect. It is not known whether these findings are secondary to cultural demands or biological differences between men and women.

In our society, the thin, athletic look in women is highly prized, whereas being "fat" is seen as unhealthy and unattractive. Under these conditions, it is no surprise that obese women feel stigmatized and depressed. In older men and women, who are less likely to be competing for sexual partners, obesity is less of a risk for depression. This would further suggest that social concerns, more than metabolic effects, are the main cause of depression in obese individuals. However, it is unlikely that depression among the obese is entirely due to social phenomenon. When being overweight includes abdominal obesity, a feature commonly related to Metabolic Syndrome, the risk of depression increases in both men and women[49]. An interesting new finding is that serum adiponectin levels are low in both Major Depression and

Metabolic Syndrome[50]. Adiponectin levels decrease as visceral adipocytes load up with fat as occurs in Metabolic Syndrome. This adipocytokine not only lowers appetite, but may also help sustain mood as well. Over all, the data suggest that when obesity is part of the larger picture of Metabolic Syndrome, depression becomes a greater risk.

If obesity, insulin resistance and Metabolic Syndrome contribute to Major Depression, then weight loss and improvement in insulin resistance and other metabolic factors should lead to improvement in depression. Reports on mood after weight loss almost uniformly show that obese individuals have less depression and anxiety, and greater sense of well-being after losing weight. Moreover, improvements in mood after weight loss are usually accompanied by improvements in insulin sensitivity, glucose tolerance, and blood lipid levels[51]. The opposite also holds true. When depressed patients with insulin resistance and poor glucose tolerance are successfully treated for depression with antidepressants, their metabolic symptoms also improve[52]. Thus, when Metabolic Syndrome, obesity and Major Depression coexist, helping one appears to help the other two conditions.

THE LIVER, METABOLIC SYNDROME AND MAJOR DEPRESSION

Although depression is a well-known side effect of interferon therapy for Hepatitis C, liver abnormalities, *per se*, are rarely considered as playing a role in the etiology of depression. However, there are many ways in which abnormal liver function in Metabolic Syndrome could cause disturbances in brain activity and mood. The liver enzyme tryptophan oxygenase is activated in gluconeogenesis. It converts tryptophan to kynurenine, and thus shifts it away from serotonin synthesis and into the pathway for making glucose. Depletion of serum tryptophan is known to exacerbate depression[53]. Interestingly, many antidepressants inhibit tryptophan oxygenase activity in rat liver[54]. Gluconeogensis is chronically disinhibited in Metabolic Syndrome, and this both increases the activity of tryptophan oxygenase in the liver and decreases serum tryptophan levels. In rats made diabetic with streptozocitin, liver tryptophan oxygenase increases by 60 percent in the liver, while brain tryptophan levels decrease by nearly a third. Serotonin turnover in the brain also decreases[55]. It is not known if insulin resistance in liver tissue leads to a similar disinhibition of tryptophan oxygenase. Cortisol, which is often elevated in Metabolic Syndrome, also stimulates gluconeogenesis in the liver, in part by stimulating tryptophan oxygenase[56]. It remains to be determined if the effects of diminished insulin activity and hypercortisolemia on tryptophan oxygenase and serum tryptophan are additive.

Whereas vagus nerve stimulation is seen as cutting edge treatment for treatment-resistant depression, it should be recalled that the liver is the origin of many of the afferent fibers that reach the brain through the vagus nerve. The vagal afferents from the liver inform the brain about energy status[57], and the liver can affect insulin sensitivity in this manner[58]. Lesioning the vagal afferents from the liver attenuates the normal hormonal response to exercise[59]. A variety of inflammatory cytokines elevated in Metabolic Syndrome can also stimulate afferent vagal activity, and are known to modulate hypothalamic function[60]. It is not known how Metabolic Syndrome alters afferent input from the liver to the brain, or if that change affects mood. It would be of interest to determine to what extent electric vagus nerve stimulation mimics the natural stimulation of the vagus nerve by a healthy liver.

CHOLESTEROL AND SUICIDE

Another phenomenon that may be generated by the relationship between Metabolic Syndrome and Major Depression is the puzzling connection between depression, cholesterol and suicide. It was noted in a 1992 publication in *The Lancet* that there was an increase in the risk of suicide and other forms of violent death in men who had been treated for hypercholesterolemia with medication and diet[61]. This possibility was supported by findings in the animal literature that reduction in cholesterol in the membranes of neurons in the brain, in turn, reduced the number of certain serotonin receptors. Reductions in serotonergic activity have been linked to depression and aggressive behavior.

Subsequent studies have generated mixed results. Some have supported the hypothesis that lowering cholesterol increased the risk of violent death, particularly by suicide. In one study, men and women in the lowest 25 percent of cholesterol levels were nearly three times as likely to commit suicide than those in the highest 25 percent of cholesterol levels[62]. However, pravastatin, which reduces serum cholesterol but does not cross the blood-brain barrier, has recently been found to have no effect on anger, depression, or anxiety[63]. To only add to the confusion, there has been a report that high levels of serum cholesterol may be related to increased risk of suicide[64].

It is possible that the connection between cholesterol and suicide may depend upon whether it is HDL or LDL that is deficient. There is evidence from several sources that links low HDL with increased risk of suicide[65]. In one study, both low HDL and serious suicide attempts were observed in the context of inflammatory signs consistent with Metabolic Syndrome[66]. At first blush, high levels of HDL almost seem to have a protective effect. In the study noted above, in which increases in total serum cholesterol were related to increases in death by violent suicide, the risk of this type of violent death decreased as the HDL component of the cholesterol increased. However, there are also studies showing no relationship between HDL and suicide[67].

HDL could indirectly affect the brain. For example, the HDL3 form of the lipoprotein appears to help catalyze the conversion of phospholipids into phosphatidylcholine, which is then actively transported across the blood-brain barrier into the neural tissue[68]. This process may serve as an important source of polyunsaturated fatty acid and, ostensibly, choline in the brain. It has been suggested that abnormalities in phosphatidylcholine activity in the brain might contribute to mood disorder[69]. Overall, however, it is difficult see how changes in serum HDL could directly affect brain activity. The brain is rich in cholesterol, and the amount of cholesterol in brain tissue can alter the way in which certain receptors perform in neuronal membranes. However, the brain does not depend on serum cholesterol as a source of the substance. Most of the cholesterol in the brain is synthesized *in situ*[70] by astrocytes and other glial cells. These cells are also able to synthesize the lipoproteins that ferry cholesterol around in brain tissue. In fact, when astrocytes are cultured, it is possible to find fully formed HDL cholesterol in the medium. If there is a relationship between HDL level and suicidal behavior in depressed patients, it is unlikely that this is due to HDL acting directly on the brain to alter that risk.

The inconsistencies among the studies of the relationship between HDL, depression, and suicide could be due to confounding by other variables such as anxiety. Comorbidity of anxiety and depression increases the risk of suicide beyond what is seen in depression alone[71]. Moreover, the comorbidity of anxiety and depression decreases HDL while increasing total cholesterol beyond what is found in either condition alone[72]. Most likely, HDL serves as a marker for some other condition, such as co-morbid anxiety, sympathetic hyperactivation, poor diet, lack of exercise or various components of Metabolic Syndrome that might independently contribute to adverse changes in brain activity and the risk of suicide.

SEASONAL AFFECTIVE DISORDER

Seasonal Affective Disorder (SAD) is a form of depression marked by depressed mood in the winter months when exposure to sunlight is decreased. Symptoms tend to remit in the spring and summer. Along with depressed mood, symptoms often include hypersomnolence, carbohydrate craving and weight gain[73]. In many respects, SAD is a seasonal variation of the so-called Atypical Depression that often affects women. The symptoms of SAD lead one to suspect that there may be a fundamental problem in energy metabolism in the condition. Moreover, SAD might share pathological features with Metabolic Syndrome.

The disturbances in energy metabolism and the seasonal nature of SAD beg us to look for pathology among the various mechanisms in the brain that serve to adapt food intake and energy metabolism to environmental demands. The food intake and energy metabolism of many mammals are strongly influenced by seasonal change. A number of hormonal systems have evolved to synchronize behavior with seasonal variation in the environment. Genes coding for components of some of those systems have been suspected of playing a role in Neel's thrifty gene theory, and, in turn, have been implicated in Metabolic Syndrome.

No clear relationship has been established between SAD and insulin resistance. However, there is a wealth of data from animal studies showing that a lack of melatonin, a substance used frequently in the treatment of SAD, exacerbates insulin resistance and causes a compensatory hyperinsulinemia[74]. Young men with Metabolic Syndrome can have perfectly normal serum levels of melatonin[75]. Thus, Metabolic Syndrome does not entail abnormal melatonin production. However, it is possible that abnormalities in melatonin production could contribute to Metabolic Syndrome.

As with other forms of depression, patients with SAD have increases in serum levels of certain inflammatory cytokines. IL-6 and soluble receptors for IL-6 and IL-2 are elevated in these patients[76]. Two weeks of light treatment did not reduce these abnormal levels of IL-6, which suggests that abnormalities in IL-6 and other cytokines may reflect a trait that predisposes patients to SAD.

Another abnormality found in patients with SAD is that serum leptin levels do not parallel their increases in weight during the winter when their symptoms are experienced[77]. Leptin is thought to help control appetite, and it acts in concert with insulin to manage fat and carbohydrate metabolism. It is yet another substance that has been implicated in Metabolic Syndrome. Decreases in leptin in the brain have recently been associated with depression and suicide in humans[78].

The melanocortins are also of interest in this context. The melanocortin system is a target of leptin, and in animal studies its activity changes with the seasons[79]. Mutations of the melanocortin 4 receptor (MC4R) have been linked to human obesity[80]. MC4R antagonists have significant antidepressant effects[81], and studies are under way to evaluate their clinical usefulness. Interestingly, there has been a case report of two sisters who each had an unusual mutation of the MC4R that eliminated its activity. Both women suffered morbid obesity and diabetes, one was diagnosed with Bipolar Affective Disorder, and the other was cyclothymic[82].

Mutations of the 5-HT2C receptor have been associated with obesity and affective disorders. It isn't clear if genetic aberrations of the 5-HT2C receptor predispose to SAD. There is, in fact, some evidence to the contrary[83]. However, it has been reported that a single amino acid substitution in the 5-HT2C receptor protein plays a significant role in determining whether or not women with SAD will suffer weight gain during their seasonal depression[84]. It is important to note that the 5-HT2C receptor may undergo posttranscriptional changes that substantially change their activity. Such changes in the 5-HT2C receptor have been identified in the prefrontal cortexes of severly depressed and schizophrenic patients[85]. A number of other substances, such as neuropeptide Y, agouti-related peptide, cocaine-and-amphetamine-regulated transcript, cholecystokinin, and others play important roles in food intake, mood, and seasonal changes in behavior. It remains to be determined if one or more of these neurohormones are factors in the etiology of SAD or its hypothetical relationship with Metabolic Syndrome.

A potentially important effect on mood may arise from a subtle change in color perception that occurs in diabetes and, possibly, in Metabolic Syndrome. When I was a graduate student at the University of British Columbia, I had the pleasure of working with a delightful, grandfatherly professor named Romuald Lakowsky. In the 1970s, he was one of the first to recognize that patients with diabetes experience significant changes in their perception of blue light[86]. This has subsequently become a well-recognized phenomenon. Dr. Lakowski suspected that these changes parallel the severity of retinal damage in diabetics. However, it has recently been found that these changes in perception of blue light may occur well before the usual evidence of retinopathy, and they may be due to early, more subtle effects of simple hyperglycemia[87]. I am not aware of any studies of color perception in Metabolic Syndrome. However, the poor glucose tolerance, inflammation and oxidative damage seen in Metabolic Syndrome might well produce deficits in color vision as has been well-established to exist in diabetes.

Dr. Norman Rosenthal's seminal work laid the foundation for using light therapy as treatment for SAD[88]. In this regard, blue light may be particularly important. Narrow band, blue-light-emitting diodes have been shown to reverse the symptoms of depression in SAD[89]. Blue light also has the unique ability to affect melatonin production and circadian rhythms[90]. It is possible that even before the onset of diabetes and retinopathy, sufferers of Metabolic Syndrome may experience subtle changes in the perception of blue light that could adversely affect circadian rhythm, sleep, and mood.

BIPOLAR AFFECTIVE DISORDER

Bipolar Affective Disorder is a psychiatric illness related to Major Depression. It used to be called "Manic Depression." However, many psychiatrists suspect that the classical Manic Depression of years ago represents only the most extreme form of a spectrum of disorders characterized by ups and downs in mood. As it became recognized that milder forms of the illness exist, "Manic Depression" was replaced by the term Bipolar Affective Disorder Type I (BPAD-I), with the milder form being referred to as Bipolar Affective Disorder Type II (BPAD-II).

BPAD-I can be an extremely severe and disabling illness. At its worst, the illness can produce hallucinations, agitation and delusions indistinguishable from the psychosis of paranoid schizophrenia. In fact it has not been uncommon for patients with BPAD-I to be misdiagnosed as having schizophrenia. The main clue in discriminating BPAD-I from schizophrenia is a history of times in which the patient has been completely free of mood and psychotic symptoms. Many people with BPAD-I are quite successful in their careers and relationships. Such success in life, even for brief periods of time, is unusual in schizophrenia.

During the "up" periods of BPAD-I, the patient may deny any need for sleep. They can be euphoric at these times. Just as often, they are loud, agitated, intrusive, and irritable. In a manic psychosis, patients can come to believe they have special powers or connections with God. They can be hypersexual, abuse alcohol or other substances, and spend outrageously. I recall a patient that spent $72 000 of his family's money on lawn equipment! During the down periods of BPAD-I, depression can be extremely severe. Suicide is more common in BPAD-I than in unipolar depression.

BPAD-II is a milder form of BPAD that is often mistaken for simple, unipolar Major Depression. Patients with this disorder usually present to psychiatrists or primary care physicians with complaints of feeling depressed. However, trials of antidepressants are often ineffective or even make them worse. Careful evaluation often reveals that such patients suffer periods of irritability, racing thoughts, anger, and difficulty sleeping. It is not unusual for them to be seen by several primary care physicians and psychiatrists before being diagnosed as having BPAD-II and not simple depression.

The prevalence of Metabolic Syndrome may be higher in individuals with BPAD than in those with Major Depression. As many as 30 percent of patients with BPAD meet criteria for Metabolic Syndrome. Nearly 50 percent of such patients are obese, 48 percent have elevated serum cholesterol, and nearly 40 percent have hypertension[91]. It is likely that the excesses in food and substance abuse in patients with BDAP, which can be even greater than that seen in Major Depression, are responsible for some of this increase in metabolic disturbance. However, some of this increased tendency towards Metabolic Syndrome may be due to a higher degree of sympathetic nervous system activation in BPAD[92]. Whereas the hyperinsulinemia of Metabolic Syndrome can cause increases in sympathetic nervous system acitivty[93], activation of the sympathetic nervous system can itself cause insulin resistance[94] and other aspects of Metabolic Syndrome. Both manic and hypomanic states are characterized by increases in sympathetic nervous system activity. Even when not showing symptoms of their illness, sufferers of BPAD have abnormally low heart rate variability[95] indicative of increased baseline levels of sympathetic activity. Interestingly, as with Major Depression, many of the medications used to treat BPAD act on substances in the brain that are also components of the intracellular insulin signaling system in neurons. This will be discussed in Chapter 5.

ANXIETY DISORDERS

Anxiety disorders, including Generalized Anxiety Disorder, Social Anxiety Disorder, Panic Disorder, Obsessive Compulsive Disorder, and Post Traumatic Stress Disorder are second only to substance abuse in being the most commonly diagnosed form of psychiatric illness. The lifetime prevalence of anxiety disorders in the United States is 25 percent[96]. However, this is likely an underestimation, as these disorders often go undiagnosed.

Anxiety disorders can themselves induce or aggravate certain components of Metabolic Syndrome. The most prominent common feature among the various anxiety disorders is sympathetic hyperactivation, which is likely to be responsible for the increased risk of cardiovascular death in patients with what used to be known simply as "anxiety neuroses"[97]. Sympathetic hyperactivation contributes to insulin resistance, hypertension, hyperglycemia, hypercholesterolemia, hypertriglyceridemia, and inflammation, all of which are factors in Metabolic Syndrome. Anxiety disorders may further increase the likelihood of Metabolic Syndrome through their tendency of progressing to include Major Depression, sleep disorder, and obesity as comorbidites[98].

There are a few reports that benzodiazepines can improve serum lipid profiles in patients with anxiety disorders[99]. Interestingly, benzodiazepine treatment has also been reported to increase HDL in patients with diabetes type II[100]. It is not clear if these improvements in serum lipids are due to alleviation of anxiety or to stimulation of mitochondrial bendiazepine receptors[101], which have no direct effect on anxiety. Patients with Generalized Anxiety Disorder (GAD) have the sympathetic hyperactivation that typifies anxiety disorders[102]. It is likely that sympathetic hyperactivation is responsible for the elevated serum cholesterol and triglyceride levels often seen in GAD[103]. GAD has been associated with increased risk of cardiovascular disease, and may be an independent risk factor for its development[104]. As many as 80 percent of GAD patients will develop comorbid Major Depression[105], and patients with both conditions are at even greater risk for developing Metabolic Syndrome.

Like other anxiety disorders, GAD is characterized by excessive worry. However, some of the specific DSM-IV criteria for the diagnosis of GAD, including irritability, a keyed up feeling, and muscle tension, are among the psychological characteristics that most strongly predispose individuals to Metabolic Syndrome[106]. Although it is not a DSM-IV criterion for diagnosis, the tendency for outbursts of anger seen in patients with GAD[107] is another risk factor for Metabolic Syndrome. Curiously, an unexpectedly high percentage of patients with NAFLD also have a history of GAD[108]. The accumulation of fat in the liver that defines NAFLD is thought to be an end result of Metabolic Syndrome. I am not aware of reports associating other anxiety disorders with NAFLD.

The primary difference between Panic Disorder and GAD is that the anxiety of Panic Disorder is less persistent, but of greater intensity during actual panic attacks. These attacks can be so severe that the patient often begins to experience anticipatory anxiety over when the next attack may occur. At this point the patient may begin to show signs and symptoms similar to those of GAD. However, the anticipatory anxiety of Panic Disorder is not as commonly characterized as having the muscle tension, restlessness, irritability, and anger that puts sufferers of GAD at higher risk of developing Metabolic Syndrome.

It is surprising that hypercortisolemia, which can contribute to Metabolic Syndrome, has not been found to be an important factor in Panic Disorder. Although persistent elevations of serum cortisol levels can be seen in severe cases of Panic Disorder[109], the primary abnormality in panic disorder is simply an exaggerated cortisol response to novel situations[110]. There is no evidence of a relationship between Panic Disorder and insulin resistance, the *sine qua non* of Metabolic Syndrome. Nonetheless, some important aspects of Metabolic Syndrome, such as increases in serum cholesterol[111], are commonly seen in Panic Disorder. Moreover, while sympathetic hyperactivity is not persistent in Panic Disorder[112], signs of hyperactivity are apparent in reports of hypertension[113] and reduced heart rate variability in this anxiety disorder[114]. There is also a relationship between Panic Disorder and obesity. These findings may explain why individuals suffering from Panic Disorder have an increased risk of cardiac death[115].

Social Anxiety Disorder is sometimes dismissed as mere shyness. This is despite the fact that sufferers of Social Anxiety Disorder have a high risk of comorbid Major Depression and suicide[116]. There is only one study of relationships between Social Anxiety Disorder and Metabolic Syndrome[117], but that study is revealing. Women diagnosed as having Social Anxiety Disorder by DSM IV criteria tend to have high serum triglycerides, elevated LDL, low HDL, and increased waist-to-hip ratios in comparison to normal controls. Those women would meet criteria for the diagnosis of Metabolic Syndrome.

In Obsessive Compulsive Disorder (OCD), the patient experiences considerable anxiety in the context of recurrent thoughts or behaviors that they cannot resist despite an ability to see them as having no basis in reality.

Our knowledge of OCD provides what is perhaps the strongest evidence of a biological basis of anxiety. It has long been known that OCD-like symptoms can arise in children after streptococcal infections, such as rheumatic fever. This is due to antistreptococcus antibodies cross-reacting with neurons in the basal ganglia[118], and the phenomenon is referred to as pediatric autoimmune neuropsychiatric disorders associated with streptococcal infections, or PANDAS[119].

Due to the associations of OCD with aberrant immunological processes, such as PANDAS, a number of researchers have looked for abnormalities in cytokines and immune cell function in the disorder. In most cases, serum levels of the cytokines IL-1 beta, IL-2 , IL-3, and IL-6 are no different in patients with OCD than in controls[120]. Curiously, there are now several reports that serum TNF-alpha levels are low in OCD[121]. This is in contrast with elevated TNF-alpha levels often seen in Major Depression and Metabolic Syndrome. As in the other anxiety disorders, OCD patients tend to have high LDL and VLDL, hypertriglyceridemia, and low serum levels of HDL[122]. Unlike the other anxiety disorders, OCD does not seem to be accompanied by decreases in heart rate variability[123] that would suggest sympathetic hyperactivity.

Post Traumatic Stress Disorder (PTSD) arises from traumatic events "in which the sufferer experiences, witnesses, or is confronted by either actual or threatened loss of life or serious injury."[124] Although sympathetic hyperactivity can be observed in all of the anxiety disorders, except perhaps OCD, it is most persistent and severe in PTSD. A puzzling difference is that, unlike sufferers of other anxiety disorders, patients with PTSD often have low serum cortisol levels. Cortisol levels remain low even when PTSD is comorbid with Major Depression, which itself is generally associated with hypercortisolemia[125].

Even without the extra adverse effects of hypercortisolemia, patients with PTSD are at increased risk for developing Metabolic Syndrome. This tendency has been observed in both combat soldiers[126] and civilian police officers[127] diagnosed with PTSD after experiencing severe, life-threatening traumas. The incidence of obesity among those veterans was described as being "strikingly" above rates in the general population, and a disproportionately large percentage of those veterans received medications to treat hypertension, diabetes mellitus, and dyslipidemia. The police officers with PTSD were approximately three times more likely than fellow officers to meet NCEP: ATP III criteria for diagnosis of Metabolic Syndrome. Those with the most severe symptoms of PTSD were most likely to suffer Metabolic Syndrome.

In considering the puzzling differences in cortisol levels among the anxiety disorders, it is worthwhile to note a paper by Drs. Charles Raison and Andrew Miller that cautions us against drawing firm and simple distinctions between the pathophysiological effects of hypo- and hypercortisolemia[128]. In this paper, they point out the existence of what is essentially cortisol resistance. As in insulin resistance, resistance to cortisol may vary in magnitude across tissues in the body. In some tissues, hypercortisolemia may eventually cause desensitization to cortisol's effects, which functionally would be identical to a lack of cortisol as occurs in PTSD. The increases in sympathetic nervous activity seen in patients with severe Major Depression[129] and hyper-cortisolemia may be partially due to disinhibition of sympathetic activity as the system becomes resistant to cortisol's dampening effects[130]. Thus, a *functional* lack of glucocorticoid control of sympathetic activity may be a common feature in both Major Depression and PTSD.

SCHIZOPHRENIA

Schizophrenia is a devastating psychiatric illness whose sufferers are also likely to develop Metabolic Syndrome. Whereas Major Depression and BPAD are diseases of emotion and activity, schizophrenia is primarily a disease of thought and perception. Common symptoms of schizophrenia include thought disorganization, delusions and hallucinations. In many cases, the thought disorganization leaves sufferers distant and detached from other human beings. Motivation slips away, and they may no longer be moved by the affairs of humanity. It is not unusual for schizophrenics to also suffer severe mood disorders. In some cases, the mood disorders arise from the terrible frustration and loss that is experienced in the progressive disintegration of the mind. However, mood and thought disorder may coexist on a persistent basis. In such cases, the illness meets criteria for the diagnosis of Schizoaffective Disorder.

It has been estimated that in the United States as many as 60 percent of people with schizophrenia meet the criteria for Metabolic Syndrome, as opposed to 30 percent for the general population. Part of the tendency for schizophrenics to develop Metabolic Syndrome is due to bad health habits, poor food choices, and lack of regular medical care. Schizophrenics tend to have higher rates of obesity. This is particularly the case in women with schizophrenia[131]. When the tendency for increases in body mass index are set aside, schizophrenics have still been found to have higher amounts of visceral fat than people without the illness[132]. Antipsychotic medication used to treat schizophrenia can cause insulin resistance and Metabolic Syndrome. I will discuss this problem in Chapter 5. However, increases in visceral fat, which typifies Metabolic Syndrome, can be observed even in patients that have never been given antipsychotic medications. There had also been reports in the literature of abnormalities in glucose tolerance years before the existence of any antipsychotic medication[133].

Schizophrenics have increased serum levels of inflammatory cytokines[134]. It is not clear if these elevated cytokine levels are an underlying component of schizophrenia itself, or if they arise because of the stress and poor health habits of schizophrenics. Nonetheless, many of these inflammatory cytokines are products of visceral fat cells, which are abundant in these individuals. Many are the same cytokines that are elevated in Metabolic Syndrome. Metabolic Syndrome predisposes its sufferers to cardiovascular disease, and schizophrenics are more likely to die of cardiovascular disease than normal individuals. Whereas one-half of people without schizophrenia die of cardiovascular disease, roughly two-thirds of schizophrenics die from this cause[135].

METABOLIC SYNDROME, SCHIZOPHRENIA, AND THE NMDA RECEPTOR

Following the introduction of chlorpromazine in the 1950s, a cadre of antipsychotic medications was developed and used for the treatment of schizophrenia. The discovery that the first effective antipsychotic drugs acted primarily as antagonists at dopamine receptors led to the hypothesis that schizophrenia was the result of overactivity of dopamine in the brain. The modern, so-called atypical antipsychotic medications also possess a significant degree of dopamine antagonism in their pharmacological profile, and the blockade of dopamine receptors seems to be a necessary component of their antipsychotic effects. Nonetheless, most psychiatrists and neuroscientists have persisted in the belief that defects in neurotransmitter systems other than dopamine are likely to contribute to the pathophysiology of schizophrenia.

The psychosis-inducing effects of hallucinogenic drugs have long provided insight into the possible neurochemical mechanisms by which the symptoms of schizophrenia arise. The drugs phencyclidine (PCP) and ketamine produce schizophrenia-like symptoms by acting as antagonists at a subtype of glutamate receptor in the brain called the N-methyl-D-aspartate (NMDA) receptor. This has led to the hypothesis that decreases in activity of the NMDA receptor might contribute to the distressing cognitive and perceptual disturbances of schizophrenic patients[136]. There is now considerable data to support this hypothesis. Quite recently, medications that act specifically on glutamate systems in the brain have been found to help alleviate some symptoms of schizophrenia. An experimental drug, LY2140023, stimulates the mGluR2/2 receptor that acts to presynaptically inhibit glutamate release in certain areas of the cortex. LY2140023 may be as effective in the treatment of schizophrenia as the dopamine antagonists that have been the mainstay of the pharmacological treatment of the illness[137].

There are several pathophysiological changes in Metabolic Syndrome that could add to existing decreases in activity at NMDA receptors and thus exacerbate the symptoms of schizophrenia. This likelihood is greater in patients whose Metabolic Syndrome has evolved to the point of including hypercortisolemia and liver dysfunction from NAFLD.

Serum levels of Asymmetrical Dimethylarginine (ADMA) are increased in both Metabolic Syndrome and schizophrenia[138]. ADMA is an endogenous inhibitor of NO synthesis[139]. Because NO is a major intracellular mediator of NMDA receptor activity, inhibition of NO synthesis antagonizes some aspects of NMDA receptor activation. Application of ADMA to hippocampal slices prevents the NMDA receptor-mediated effect of long-term potentiation[140]. ADMA also competes for uptake into cells with arginine, the amino acid that serves as the substrate for synthesis of NO in the brain and elsewhere in the body[141]. ADMA might itself also be taken up into the brain by this mechanism. Thus, ADMA might contribute both directly and indirectly to decreases in NMDA receptor activity in the brain.

The increases in serum ADMA may be due in part to homocysteine, which is also elevated in Metabolic Syndrome and schizophrenia. Homocysteine increases ADMA by preventing its destruction by the enzyme dimethylarginine dimethylaminohydrolase. Homocysteine may have further adverse impact on the NMDA receptor by stimulating the production of the NMDA receptor antagonist kynurenine in the endothelium[142]. Metabolic Syndrome may increase kynurenine levels by other mechanisms as well. Whereas insulin inhibits, cortisol stimulates the enzyme tryptophan oxygenase that converts tryptophan to kynurenine. It is not known what the state of insulin resistance might contribute to that process, although it might be expected to further disinhibit tryptophan oxygenase activity.

The methyl groups used to synthesize ADMA are donated by s-adenosylmethionine (SAMe). I offer a more thorough discussion of SAMe in Chapter 10. It is not known if increases in ADMA synthesis in Metabolic Syndrome result in a concomitant depletion of SAMe. However, when Metabolic Syndrome has progressed, as it sometimes does, to include NAFLD, then serum SAMe levels would begin to decline. The significance of SAMe in the context of the NMDA receptor is that SAMe is required for synthesis of the polyamines in the brain. The polyamines, which include putrescine, spermine, and spermidine, enhance the binding of glycine to the NMDA receptor complex and increase the likelihood of NMDA receptor activation. Thus, depletion of SAMe could contribute to decreases in NMDA receptor activity that, in turn, may increase the psychotic symptoms of schizophrenia.

FIBROMYALGIA

Fibromyalgia is a puzzling disorder. It is generally described as "hurting all over". Although the American College of Rheumatology long ago established a list of diagnostic criteria[143], there are still many physicians who refuse to believe that fibromyalgia exists. Many doctors see it as a form of depression, anxiety, personality disorder, or outright malingering. This is why many psychiatrists become so familiar with fibromyalgia. Such patients are sent to us by doctors who assume it is all in their head. The illness is characterized by severe, diffuse, and chronic musculoskeletal pain. However, it is also often accompanied by fatigue, changes in mood, and cognitive function, and irritable bowel symptoms. In all fairness to the doctors who simply throw up their hands, the range and complexity of the symptoms can be bewildering.

It is tempting to draw a connection between fibromyalgia and Metabolic Syndrome. Several reports associate fibromyalgia with obesity[144] and diabetes type II[145]. Moreover, in individuals suffering fibromyalgia, there is increased incidence of both insulin resistance[146] and elevated fasting serum insulin levels[147], which are both hallmarks of Metabolic Syndrome. Serum levels of cytokines that are elevated in Metabolic Syndrome, including IL-8[148] and IL-6[149], are also elevated in fibromyalgia. Weight loss improves both fibromyalgia[150] and Metabolic Syndrome.

The above findings would suggest a relationship between fibromyalgia and Metabolic Syndrome. However, some cytokines that are almost always found to be elevated in Metabolic Syndrome, including C-reactive protein, IL-1-beta, and TNF-alpha, are usually unchanged in fibromyalgia. Moreover, there are apparent inconsistencies in the literature concerning hyperlipidemia and hypercortisolemia in fibromyalgia. In Metabolic Syndrome cholesterol levels and serum triglycerides are elevated, and serum cortisol is often increased as well. In one report, no changes in the serum lipids were observed in patients with fibromyalgia[151], and cortisol levels have been described as being normal or even mildly insufficient in fibromyalgia[152]. However, a recent study has found that both triglycerides and LDL cholesterol are elevated, and serum cortisol signifcantly increased in sufferers of fibromyalgia[153]. The conclusion of that study is that women with fibrolmyalgia are nearly six times as likely as women without the illness to meet criteria for diagnosis of Metabolic Syndrome.

POLYCYSTIC OVARY DISEASE

Another illness with suggestions of connections between psychiatric illness and Metabolic Syndrome is Polycystic Ovary Disease (PCO). This is a surprisingly common condition that may affect as many as 5–10 percent of American women[154]. As many as 50 percent of women with PCO[155] are also clinically depressed. The prevalence of sleep disorders among women with PCO may reach as high as 90 percent. These women are particularly prone to sleep apnea, as well as having difficulty getting and staying asleep[156]. In Chapter 8, I will discuss in detail the relationships between sleep disorders, Metabolic Syndrome, and psychiatric disorders.

The primary symptoms of PCO are irregular menstrual periods, failure of ovulation, infertility, obesity, and increases in facial and body hair. Laboratory studies of PCO reveal high serum levels of testosterone and estrogen. One of the most striking characteristics of PCO is insulin resistance. No one is certain what causes the insulin resistance in PCO. Some suspect that high levels of testosterone in PCO cause increases in abdominal fat deposition. The cytokines released from visceral adipocytes might, in turn, contribute to insulin resistance and other aspects of Metabolic Syndrome. However, there are theories that insulin resistance is the first step in the development of PCO. Insulin resistance induces a compensatory increase in serum insulin levels. When present in high concentration, insulin can act in the ovaries at receptors for the structurally similar hormone, insulin-like growth factor. Stimulation of insulin-like growth factor receptors results in overproduction of testosterone by the ovary. The increases in testosterone might then produce many of the symptoms of PCO. Various treatments can be used to stop the production of testosterone by the ovaries. However, doing so does not reverse the insulin resistance of PCO. This would tend to support the notion that insulin resistance is what initiates the progression of PCO.

REFERENCES

1. Gavard, J.A., Lustman, P.J., and Clouse, R.E., Prevalence of depression in adults with diabetes. An epidemiological evaluation. *Diabetes Care* 1993; 16:1167–1178.
2. Hess, Z. et al., Metabolic syndrome and latent depression in the population sample. *Cas. Lek. Cesk.* 2004; 143:840–844.
3. Everson-Rose, S.A. et al., Depressive symptoms, insulin resistance, and risk of diabetes in women at midlife. *Diabetes Care* 2004; 27:2856–2862.

4. Petrlová, B. et al., Depressive disorders and the metabolic syndrome of insulin resistance. *Semin. Vasc. Med.* 2004; 4:161–165.

5. Räikkönen, K., Matthews, K.A., and Kuller, L.H., The relationship between psychological risk attributes and the metabolic syndrome in healthy women: antecedent or consequence? *Metabolism* 2002; 51:1573–1577.

6. Vitaliano, P.P. et al., A path model of chronic stress, the metabolic syndrome, and coronary heart disease. *Psychosom. Med.* 2002; 64:418–435.

7. Bonnet, F. et al., Depressive symptoms are associated with unhealthy lifestyles in hypertensive patients with the metabolic syndrome. *J. Hypertens.* 2005; 23:611–617.

8. Posternak, M.A. and Zimmerman, M., Symptoms of atypical depression. *Psychiatry Res.* 2001; 104:175–181.

9. Kinder, L.S. et al., Depression and the metabolic syndrome in young adults: findings from the Third National Health and Nutrition Examination Survey. *Psychosom. Med.* 2004; 66: 316–322.

10. Rasgon, N.L. and Kenna, H.A., Insulin resistance in depressive disorders and Alzheimer's Disease: revisiting the missing link hypothesis. *Neurobiol. Aging* 2005; 26 (Suppl 1):103–107.

11. Craft, S. and Watson, G.S., Insulin and neurodegenerative disease: shared and specific mechanisms. *Lancet Neurol.* 2004; 3:169–178.

12. Bingham, E.M. et al., The role of insulin in human brain glucose metabolism: an 18fluoro-deoxyglucose positron emission tomography study. SA. *Diabetes* 2002; 51:3384–3390.

13. Tschritter, O. et al., The cerebrocortical response to hyperinsulinemia is reduced in overweight humans: a magnetoencephalographic study. *PNAS* 2006; 103:12103–12108.

14. Anthony, K. et al., Attenuation of insulin-evoked responses in brain networks controlling appetite and reward in insulin resistance. *Diabetes* 2006; 55:2986–2992.

15. Mielke, J.G. et al., A biochemical and functional characterization of diet-induced brain insulin resistance. *J. Neurochem.* 2005; 93:1568–1578.

16. Davidson, J.R. et al., Effectiveness of chromium in atypical depression: a placebo-controlled trial. *Biol. Psychiatry* 2003; 53:261–264.

17. McLeod, M.N., Gaynes, B.N., and Golden, R.N., Chromium potentiation of antidepressant pharmacotherapy for dysthymic disorder in 5 patients. *J. Clin. Psychiatry* 1999; 60:237–240.

18. Docherty, J.P. et al., A double-blind, placebo-controlled, exploratory trial of chromium picolinate in atypical depression: effect on carbohydrate craving. *J. Psychiatr. Pract.* 2005; 11:302–314.

19. Meijer, L., Flajolet, M., and Greengard, P., Pharmacological inhibitors of glycogen synthase kinase 3. *Trends. Pharmacol. Sci.* 2004; 25:471–480.

20. Kirshenboim, N. et al., Lithium-mediated phosphorylation of glycogen synthase kinase-3beta involves PI3 kinase-dependent activation of protein kinase C-alpha. *J. Mol. Neurosci.* 2004; 24:237–245.

21. Su, K.P. et al., Omega-3 fatty acids in major depressive disorder. A preliminary double-blind, placebo-controlled trial. *Eur. Neuropsychopharmacol.* 2003; 13:267–271.

22. Delarue, J. et al., N-3 long chain polyunsaturated fatty acids: a nutritional tool to prevent insulin resistance associated to type 2 diabetes and obesity? *Reprod. Nutr. Dev.* 2004; 44: 289–299.

23. McEwen, B.S., Protection and damage from acute and chronic stress: allostasis and allostatic overload and relevance to the pathophysiology of psychiatric disorders. *Ann. N.Y. Acad. Sci.* 2004; 1032:1–7.

24. Alfonso, J. et al., Identification of genes regulated by chronic psychosocial stress and antidepressant treatment in the hippocampus. *Eur. J. Neurosci.* 2004; 19:659–666.

25. Nemeroff, C.B., Clinical significance of psychoneuroendocrinology in psychiatry: focus on the thyroid and adrenal. *J. Clin. Psychiatry* 1989; 50 (Suppl):13–20.

26. Mendelson, S.D. and McEwen, B.S., Autoradiographic analyses of the effects of adrenalectomy and corticosterone on 5-HT1A and 5-HT1B receptors in the dorsal hippocampus and cortex of the rat. *Neuroendocrinology* 1992; 55:444–450.

27. Brown, G.L. and Linnoila, M.I., CSF serotonin metabolite (5-HIAA) studies in depression, impulsivity, and violence. *J. Clin. Psychiatry* 1990; 51 (Suppl):31–41.

28. Pitchot, W. et al., 5-Hydroxytryptamine 1A receptors, major depression, and suicidal behavior. *Biol. Psychiatry* 2005; 58:854–858.

29. Zahorodna, A. and Hess, G., Imipramine and citalopram reverse corticosterone-induced alterations in the effects of the activation of 5-HT(1A) and 5-HT(2) receptors in rat frontal cortex. *J. Physiol. Pharmacol.* 2006; 57:389–399.

30. Schatzberg, A.F. et al., A corticosteroid/dopamine hypothesis for psychotic depression and related states. *J. Psychiatr. Res.* 1985; 19(1):57–64.

31. Belanoff, J.K. et al., Cortisol activity and cognitive changes in psychotic major depression. *Am. J. Psychiatry* 2001; 158:1612–1616.

32. Friedman, T.C. et al., Carbohydrate and lipid metabolism in endogenous hypercortisolism: shared features with metabolic syndrome X and NIDDM. *Endocr. J.* 1996; 43:645–655.

33. Rizza, R.A., Mandarino, L.J., and Gerich J.E., Cortisol-induced insulin resistance in man: impaired suppression of glucose production and stimulation of glucose utilization due to a postreceptor detect of insulin action. *J. Clin. Endocrinol. Metab.* 1982; 54:131–138.

34. Weber-Hamann, B. et al., Hypercortisolemic depression is associated with increased intra-abdominal fat. *Psychosom. Med.* 2002; 64:274–277.

35. van Rossum, E.F. et al., Polymorphisms of the glucocorticoid receptor gene and major depression. *Biol. Psychiatry* 2006; 59:681–688.

36. Rosmond, R. et al., A glucocorticoid receptor gene marker is associated with abdominal obesity, leptin, and dysregulation of the hypothalamic-pituitary-adrenal axis. *Obes. Res.* 2000; 8:211–218.

37. Sluzewska, A. et al., Indicators of immune activation in major depression. *Psychiatry Res.* 1996; 64:161–167.

38. Silverman, M.N. et al., Immune modulation of the hypothalamic-pituitary-adrenal (HPA) axis during viral infection. *Viral. Immunol.* 2005; 18:41–78.

39. Pace, T.W. et al., Increased stress-induced inflammatory responses in male patients with major depression and increased early life stress. *Am. J. Psychiatry* 2006; 163:1630–1633.

40. McKay, L.I. and Cidlowski, J.A., Molecular control of immune/inflammatory responses: Interactions between nuclear factor-kappa B and stweroid receptor-signaling pathways. *Endocr. Rev.* 1999; 20:435–459.

41. Raison, C.L., Capuron, L., and Miller, A.H., Cytokines sing the blues: inflammation and the pathogenesis of depression. *Trends Immunol.* 2006; 27:25–31.

42. Kronfol, Z. and Remick, D.G., Cytokines and the brain: implications for clinical psychiatry. *Am. J. Psychiatry* 2000; 157:683–694.

43. Raison, C.L. et al., Neuropsychiatric adverse effects of interferon-alpha: recognition and management. *CNS Drugs* 2005; 19:105–123.

44. Maes, M., The immunoregulatory effects of antidepressants. *Hum. Psychopharmacol.* 2001; 16:95–103.

45. Wilkes, S., Bupropion. *Drugs Today (Barc.)* 2006; 42:671–681.

46. Tyring, S. et al., Etanercept and clinical outcomes, fatigue, and depression in psoriasis: double-blind placebo-controlled randomised phase III trial. *Lancet* 2006; 367:29–35.

47. Kaufman, K.R., Atanercept, anticytokines and mania. *Int. Clin. Psychopharmacol.* 2005; 20:239–241.

48. Heo M, et al., Depressive mood and obesity in US adults: comparison and moderation by sex, age, and race. *Int J Obes (Lond).* 2006; 30:51351–9.

49. McElroy, S.L. et al., Are mood disorders and obesity related? A review for the mental health professional. *J. Clin. Psychiatry* 2004; 65:634–651.

50. Leo, R. et al., Decreased plasma adiponectin concentration in major depression. *Neurosci. Lett.* 2006; 407:211–213.

51. Dixon, J.B. et al., Sustained weight loss in obese subjects has benefits that are independent of attained weight. *Obes. Res.* 2004; 12:1895–1902.

52. Okamura, F. et al., Insulin resistance in patients with depression and its changes during the clinical course of depression: minimal model analysis. *Metabolism* 2000; 49:1255–1260.

53. Neumeister, A., Tryptophan depletion, serotonin, and depression: where do we stand? *Psychopharmacol. Bull.* 2003; 37:99–115.

54. Badawy, A.A. and Evans, M., Inhibition of rat liver tryptophan pyrrolase activity and elevation of brain tryptophan concentration by acute administration of small doses of antidepressants. *Br. J. Pharmacol.* 1982; 77:59–67.

55. Sadler, E., Weiner, M., and Buterbaugh, G.G., Effect of streptozotocin-induced diabetes on tryptophan oxygenase activity and brain tryptophan levels in rats. *Res. Commun. Chem. Pathol. Pharmacol.* 1983; 42:37–50.

56. Green, A.R. and Curzon, G., Decrease of 5-hydroxytryptamine in the brain provoked by hydrocortisone and its prevention by allopurinol. *Nature* 1968; 220:1095–1097.

57. Randich, A. et al., Jejunal or portal vein infusions of lipids increase hepatic vagal afferent activity. *Neuroreport* 2001; 12:3101–3105.

58. Uno, K. et al., Neuronal pathway from the liver modulates energy expenditure and systemic insulin sensitivity. *Science* 2006; 312:1656–1659.

59. Lavoie, J.M., The contribution of afferent signals from the liver to metabolic regulation during exercise. *Can. J. Physiol. Pharmacol.* 2002; 80:1035–1044.

60. Borovikova, L.V. et al., Vagus nerve stimulation attenuates the systemic inflammatory response to endotoxin. *Nature* 2000; 405:458–462.

61. Engelberg, H., Low serum cholesterol and suicide. *Lancet* 1992; 339:727–729.

62. Lindberg, G. et al., Low serum cholesterol concentration and short term mortality from injuries in men and women. *BMJ* 1992; 305:277–279.

63. Stewart, R.A. et al., Long-term assessment of psychological well-being in a randomized placebo-controlled trial of cholesterol reduction with pravastatin. The LIPID Study Investigators. *Arch. Intern. Med.* 2000; 160:3144–3152.

64. Tanskanen, A. et al., High serum cholesterol and risk of suicide. *Am. J. Psychiatry* 2000; 157:648–650.

65. Zhang, J. et al., Low HDL cholesterol is associated with suicide attempt among young healthy women: the Third National Health and Nutrition Examination Survey. *J. Affect. Disord.* 2005; 89:25–33.

66. Maes, M. et al., Lower serum high-density lipoprotein cholesterol (HDL-C) in major depression and in depressed men with serious suicidal attempts: relationship with immune-inflammatory markers. *Acta Psychiatr. Scand.* 1997; 95:212–221.

67. Deisenhammer, E.A. et al., No evidence for an association between serum cholesterol and the course of depression and suicidality. *Psychiatry Res.* 2004; 121:253–261.

68. Magret, V. et al., Entry of polyunsaturated fatty acids into the brain: evidence that high-density lipoprotein-induced methylation of phosphatidylethanolamine and phospholipase A2 are involved. *Biochem. J.* 1996; 316:805–811.

69. Moore, C.M., Choline, myo-inositol and mood in bipolar disorder: a proton magnetic resonance spectroscopic imaging study of the anterior cingulate cortex. *Bipolar Disord.* 2000; 2:207–216.

70. Dietschy, M.J. and Turley, D.S., Cholesterol metabolism in the brain. *Curr. Opin. Lipidol.* 2001; 12:105–112.

71. Schaffer, A. et al., Suicidal ideation in major depression: sex differences and impact of comorbid anxiety. *Can. J. Psychiatry* 2000; 45:822–826.

72. Sevincok, L., Buyukozturk, A., and Dereboy, F., Serum lipid concentrations in patients with comorbid generalized anxiety disorder and major depressive disorder. *Can. J. Psychiatry* 2001; 46:68–71.

73. Magnusson, A. and Partonen, T., The diagnosis, symptomatology, and epidemiology of seasonal affective disorder. *CNS Spectr.* 2005; 10:625–634.

74. Nishida, S. et al., Effect of pinealectomy on plasma levels of insulin and leptin and on hepatic lipids in type 2 diabetic rats. *J. Pineal. Res.* 2003; 35:251–256.

75. Robeva, R. et al., Low testosterone levels and unimpaired melatonin secretion in young males with metabolic syndrome. *Andrologia* 2006; 38:216–220.

76. Leu, S.J. et al., Immune-inflammatory markers in patients with seasonal affective disorder: effects of light therapy. *J. Affect. Disord.* 2001; 63:27–34.

77. Cizza, G. et al., Plasma leptin in men and women with seasonal affective disorder and in healthy matched controls. *Horm. Metab. Res.* 2005; 37:45–48.

78. Eikelis, N. et al., Reduced brain leptin in patients with major depressive disorder and in suicide victims. *Mol. Psychiatry* 2006; 11:800–801.

79. Adam, C.L. and Mercer J.G., Appetite regulation and seasonality: implications for obesity. *Proc. Nutr. Soc.* 2004; 63:413–419.

80. Adan, R.A. et al., The MC4 receptor and control of appetite. *Br. J. Pharmacol.* 2006; 149:815–27.

81. Chaki, S. and Okuyama, S., Involvement of melanocortin-4 receptor in anxiety and depression. *Peptides* 2005; 26:1952–1964.

82. Mergen, M. et al., A novel melanocortin 4 receptor (MC4R) gene mutation associated with morbid obesity. *J. Clin. Endocrinol. Metab.* 2001; 86:3448.

83. Johansson, C. et al., Seasonal affective disorder and serotonin-related polymorphisms. *Neurobiol. Dis.* 2001; 8:351–357.

84. Praschak-Rieder, N. et al., A Cys 23-Ser 23 substitution in the 5-HT(2C) receptor gene influences body weight regulation in females with seasonal affective disorder: an Austrian-Canadian collaborative study. *J. Psychiatr. Res.* 2005; 39:561–567.

85. Schmauss, C., Serotonin 2C receptors: suicide, serotonin, and runaway RNA editing. *Neuroscientist* 2003; 9:237–242.

86. Kinnear, P.R., Aspinall, P.A., and Lakowski, R., The diabetic eye and colour vision. *Trans. Ophthalmol. Soc. U.K.* 1972; 92:69–78.

87. Afrashi, F. et al., Blue-on-yellow perimetry versus achromatic perimetry in type 1 diabetes patients without retinopathy. *Diabet. Res. Clin. Pract.* 2003; 61:7–11.

88. Rosenthal, N.E. et al., Seasonal affective disorder and phototherapy. *Ann. N.Y. Acad. Sci.* 1985; 453:260–269.

89. Glickman, G. et al., Light therapy for seasonal affective disorder with blue narrow-band light-emitting diodes (LEDs). *Biol, Psychiatry* 2006; 59:502–507.

90. Munch, M. et al., Wavelength-dependent effects of evening light exposure on sleep architecture and sleep EEG power density in men. *Am. J. Physiol. Regul. Integr. Comp. Physiol.* 2006; 290:R1421–1428.

91. Fagiolini, A. et al., Metabolic syndrome in bipolar disorder: findings from the Bipolar Disorder Center for Pennsylvanians. *Bipolar Disord.* 2005; 7:424–430.

92. Swann, A.C. et al., Mania: sympathoadrenal function and clinical state. *Psychiatry Res.* 1991; 37:195–205.

93. Scherrer, U. and Sartori, C., Insulin as a vascular and sympathoexcitatory hormone implications for blood pressure regulation, insulin sensitivity, and cardiovascular morbidity. *Circulation* 1997; 96:4104–4113.

94. Jamerson, K.A. et al., Reflex sympathetic activation induces acute insulin resistance in the human forearm. *Hypertension* 1993; 21:618–623.

95. Cohen, H. et al., Impaired heart rate variability in euthymic bipolar patients. *Bipolar Disord.* 2003; 5:138–143.

96. Kessler, R.C. et al., Lifetime and 12-month prevalence of DSM-III-R psychiatric disorders in the United States: results from the National Comorbidity Survey. *Arch. Gen. Psychiatry* 1994; 51:8–19.

97. Coryell, W., Noyes, R., and House, J.D., Mortality among outpatients with anxiety disorders. *Am. J. Psychiatry* 1986; 143:508–510.

98. Rosenberger, P.H., Henderson, K.E., and Grilo, C.M., Psychiatric disorder comorbidity and association with eating disorders in bariatric surgery patients: a cross-sectional study using structured interview-based diagnosis. *J. Clin. Psychiatry* 2006; 67:1080–1085.

99. Shioiri, T. et al., Effect of pharmacotherapy on serum cholesterol levels in patients with panic disorder. *Acta Psychiatr. Scand.* 1996; 93:164–167.

100. Okada, S. et al., Effect of an anxiolytic on lipid profile in non-insulin-dependent diabetes mellitus. *J. Int. Med. Res.* 1994; 22:338–342.

101. Krueger, K.E., Peripheral-type benzodiazepine receptors: a second site of action for benzodiazepines. *Neuropsychopharmacology* 1991; 4:237–244.

102. Roth, W.T., Sympathetic activation in broadly defined generalized anxiety disorder. *J. Psychiatr. Res.* 2007 (Epublished ahead of print).

103. Kuczmierczyk, A.R. et al., Serum cholesterol levels in patients with generalized anxiety disorder (GAD) and with GAD and comorbid major depression. *Can. J. Psychiatry* 1996; 41:465–468.

104. Barger, S.D. and Sydeman, S.J., Does generalized anxiety disorder predict coronary heart disease risk factors independently of major depressive disorder? *J. Affect. Disord.* 2005; 88:87–91.

105. Bakish, D., The patient with comorbid depression and anxiety: the unmet need. *J. Clin. Psychiatry* 1999; 60(Suppl. 6):20–24.

106. Raikkonen, K., Matthews, K.A., and Kuller, L.H., The relationship between psychological risk attributes and the metabolic syndrome in healthy: antecedent or consequence? *Metabolism* 2002; 51:1573–1577.
107. Rettew, D.C. et al., Exploring the boundary between temperament and generalized anxiety disorder: a receiver operating characteristic analysis. *J. Anxiety Disord.* 2006; 20:931–945.
108. Elwing, J.E. et al., Depression, anxiety, and non-alcoholic steatohepatitis. *Psychosom. Med.* 2006; 68:563–569.
109. Bandelow, B. et al., Diurnal variation of cortisol in panic disorder. *Psychiatry Res.* 2000; 95:245–250.
110. Abelson, J.L. et al., HPA axis activity in patients with panic disorder: review and synthesis of four studies. *Depress Anxiety* 2007; 24:66–76.
111. Hayward, C. et al., Plasma lipid levels inpatients with panic disorder or agorophobia. *Am. J. Psychiatry* 1989; 146:917–919.
112. Stein, M.B. and Asmundson, G.J., Autonomic function in panic disorder: cardiorespiratory and plasma catecholamine responsivity to multiple challenges of the autonomic nervous system. *Biol. Psychiatry* 1994; 36:548–558.
113. Davies, S.J. et al., Panic disorder, anxiety and depression in resistant hypertension – a case-control study. *J. Hypertens.* 1997; 15:1077–1082.
114. Katerndahl, D., Panic & plaques: panic disorder and coronary artery disease in patients with chest pain. *J. Am. Board Fam. Pract.* 2004; 17:114–126.
115. Gomez-Caminero, A., Does panic disorder increase the risk of coronary heart disease? A cohort study of a national managed care database. *Psychosom. Med.* 2005; 67:688–691.
116. Stein, M.B., An epidemiologic perspective on social anxiety disorder. *J. Clin. Psychiatry* 2006; 67(Suppl. 12):3–8.
117. Landen, M. et al., Dyslipidemia and high waist-hip ratio in women with self-reported social anxiety. *Psychoneuroendocrinology* 2004; 29:1037–1046.
118. Dale RC, et al., Incidence of anti-brain antibodies in children with obsessive-compulsive disorder. *Br. J. Psychiatry* 2005; 187:314–319.
119. Snider, L.A. and Swedo, S.E., PANDAS: current status and directions for research. *Mol. Psychiatry* 2004; 9:900–907.
120. Weizman, R. et al., Cytokine production in obsessive-compulsive disorder. *Biol. Psychiatry* 1997; 42:1187–1188.
121. Denys, D. et al., Decreased TNF-alpha and NK activity in obsessive-compulsive disorder. *Psychoneuroendocrinology* 2004; 29:945–952.
122. Agargun, M.Y. et al., Serum lipid concentrations in obsessive-compulsive disorder patients with and without panic attacks. *Can. J. Psychaitry* 2004; 49:776–778.
123. Slaap, B.R. et al., Five-minute recordings of heart rate variability in obsessive-compulsive disorder, panic disorder and healthy volunteers. *J. Affect. Disord.* 2004; 78:141–148.
124. Yehuda, R., Risk and resilience in posttraumatic stress disorder. *J. Clin. Psychiatry* 2004; 65(Suppl. 1):29–36.
125. Oquendo, M.A. et al., Lower cortisol levels in depressed patients with comorbid posttraumatic stress disorder. *Neuropsychopharmacology* 2003; 28:591–598.
126. Vieweg, W.V.R. et al., Posttraumatic stress disorder in male military veterans with comorbid overweight and obesity: psychotropic, antihypertensive, and metabolic medications. *Prim. Care Companion J. Clin. Psychiatry* 2006; 8:25–31.
127. Violanti, J.M. et al., Police trauma and cardiovascular disease: association between PTSD symptoms and metabolic syndrome. *Int. J. Emerg. Ment. Health.* 2006; 8:227–237.
128. Raison, C.L. and Miller, A.H., When not enough is too much: the role of insufficient glucocorticoid signaling in the pathophysiology of stress-related disorders. *Am. J. Psychiatry.* 2003; 160:1554–1565.
129. Maes, M. et al., Sleep disorders and anxiety as symptom profiles of sympathoadrenal system hyperactivity in major depression. *J. Affect. Disord.* 1993; 27:197–207.
130. Pace, T.W., Hu, F., and Miller, A.H., Cytokine-effects on glucocorticoid receptor function: relevance to glucocorticoid resistance and the pathophysiology and treatment of major depression. *Brain Behav. Immun.* 2007; 21:9–19.
131. Allison, D.B. et al., The distribution of body mass index among individuals with and without schizophrenia. *J. Clin. Psychiatry* 1999; 60:215–220.

132. Thakore, J.H. et al., Increased visceral fat distribution in drug-naive and drug-free patients with schizophrenia. *Int. J. Obes. Relat. Metab. Disord.* 2002; 26:137–141.

133. Lieberman, J., Metabolic changes associated with antipsychotic use. *Prim Care Companion J. Clin. Psychiatry.* 2004; 6(Suppl. 2):8613.

134. Altamura, A.C., Boin, F., and Maes, M., HPA axis and cytokines dysregulation in schizophrenia: potential implications for the antipsychotic treatment. *Eur. Neuropsychopharmacol.* 1999; 10:1–4.

135. Hennekens, C.H. et al., Schizophrenia and increased risks of cardiovascular disease. *Am. Heart J.* 2005; 150:1115–1121.

136. Coyle, J.T. Glutamate and schizophrenia: beyond the dopamine hypothesis. *Cell. Mol. Neurobiol.* 2006; 26:365–384.

137. Personal communication from Dr. John Krystal, Department of Psychiatry, Yale University.

138. Das, I. et al., Elevated endogenous nitric oxide synthase inhibitor in schizophrenic plasma may reflect abnormalities in brain nitric oxide production. *Neurosci. Lett.* 1996; 215:209–211.

139. MacAllister, R.J. et al., Metabolism of methylarginines by human vasculature; implications for the regulation of nitric oxide synthesis. *Br. J. Pharmacol.* 1994; 112:43–48.

140. Musleh, W., Yaghoubi, S., and Baudry, M., Effects of a nitric oxide synthase inhibitor on NMDA receptor function in organotypic hippocampal cultures. *Brain Res.* 1997; 770:298–301.

141. Schmidt, K., Klatt, P., and Mayer, B., Characterization of endothelial cell amino acid transport systems involved in the actions of nitric oxide synthase inhibitors. *Mol. Pharmacol.* 1993; 44:615–621.

142. Stazka, J., Luchowski, P., and Urbanska, E.M., Homocysteine, a risk factor for atherosclerosis, biphasically changes the endothelial production of kynurenic acid. *Eur. J. Pharmacol.* 2005; 517:217–223.

143. Wolfe, F. et al., The American College of Rheumatology 1990 Criteria for the Classification of Fibromyalgia. Report of the Multicenter Criteria Committee. *Arthritis Rheum.* 1990; 33:160–172.

144. Patucchi, E. et al., Prevalence of fibromyalgia in diabetes mellitus and obesity. *Recenti. Prog. Med.* 2003; 94:163–165.

145. Tishler, M. et al., Fibromyalgia in diabetes mellitus. *Rheumatol. Int.* 2003; 23:171–173.

146. Shipton, E., Controversirs, New risk factors and the neuroendocrine approach to fibromyalgia: can therapeutic progress be made at last? *NZMJ* 2003; 116:1–5.

147. Denko, C.W. and Malemud, C.J., Serum growth hormone and insulin but not insulin-like growth factor-1 levels are elevated in patients with fibromyalgia. *Rheumatol. Int.* 2005; 25:146–151.

148. Gur, A. et al., Cytokines and depression in cases with fibromyalgia. *J. Rheumatol.* 2002; 29:358–361.

149. Wallace, D. et al., Cytokines play an aetiopathogenetic role in fibromyalgia: a pilot study. *Rheumatology* 2001; 40:743–749.

150. Shapiro, J.R., Anderson, D.A., and Danoff-Burg, S., A pilot study of the effects of behavioral weight loss treatment on fibromyalgia symptoms. *J. Psychosom. Res.* 2005; 59:275–282.

151. Gurer, G. et al., Serum lipid profile in fibromyalgia women. *Clin. Rheumatol.* 2006; 25:300–303.

152. Griep, E.N., Boersma, J.W., and de Kloet, E.R., Altered reactivity of the hypothalamic-pituitary-adrenal axis in the primary fibromyalgia syndrome. *J. Rheumatol.* 1993; 20:469–474.

153. Loevinger, B.L. et al., Metabolic syndrome in women with chronic pain. *Metabolism* 2007; 56:87–93.

154. Cenk Sayin, N. et al., Insulin resistance and lipid profile in women with polycystic appearing ovaries: implications with regard to polycystic ovary syndrome. *Gynecol. Endocrinol.* 2003; 17:387–396.

155. Rasgon, N.L. et al., Depression in women with polycystic ovary syndrome: clinical and biochemical correlates. *J. Affect. Disord.* 2003; 74:299–304.

156. Tasali, E., Van Cauter, E., and Ehrmann, D.A., Relationships between sleep disordered breathing and glucose metabolism in polycystic ovary syndrome. *J. Clin. Endocrinol. Metab.* 2006; 91:36–42.

5

PSYCHIATRIC MEDICATIONS AND METABOLIC SYNDROME

Hippocrates admonished us to, "Do no harm." However, no medication is free of side effects, and many have the potential to do serious harm. Even aside from overdoses and errors in administration, the number of patients that die or are disabled by adverse reactions to medications in our nation's hospitals every year is staggering[1]. After making a diagnosis, balancing the risks and benefits of pharmacological treatments is the most challenging and perilous task we face.

Psychiatric medications are particularly prone to causing disturbances in appetite and weight gain. Some of the more recently developed medications have also been reported to cause changes in serum glucose and response to insulin that are consistent with Metabolic Syndrome. In some cases, significant morbidity and death has occurred.

ANTIDEPRESSANTS

Although Major Depression can lead to Metabolic Syndrome, some of the medications used to treat depression can also cause weight gain, changes in glucose metabolism, and adverse changes in serum lipids. The first antidepressants were the monoamine oxidase inhibitors (MAOIs). Monoamine oxidase destroys the neurotransmitters serotonin, norepinephrine, and dopamine in the brain. Blocking the enzyme prevents destruction of those neurotransmitters, and this is thought to be the mechanism by which MAOIs improve mood. Although one MAOI, phenelzine, can cause weight gain, this effect is not typical of MAOIs[2]. Most MAOIs do not directly increase appetite or weight. There are scattered reports that MAOIs may actually enhance insulin sensitivity, which could conceivably help prevent or reverse Metabolic Syndrome[3]. However, with the exception of phenelzine, MAOIs appear to have little effect on Metabolic Syndrome. The new, transdermal form of the MAOI selegiline, which is selective but not specific for MAO-B, does not appear to affect weight in patients being treated for depression[4].

The tricyclic antidepressants (TCAs) came into use with the introduction of imipramine in 1957. It was soon discovered that one of the side effects of TCAs was weight gain. This increase in weight is primarily due to increased appetite and carbohydrate craving, that are likely the result of blocking histamine and 5-HT2 receptors in the brain. TCAs can also raise serum cholesterol and triglycerides[5]; however, it is not clear if those effects are secondary to the weight gain. The worst offender is amitriptyline, whereas the most benign is desipramine. These differences are likely due to desipramine's lower affinity for histamine and 5-HT2 receptors.

When the serotonin selective reuptake inhibitors (SSRIs) came into use with the marketing of fluoxetine in 1988, psychiatrists heaved a collective sigh of relief. This was not so much for the effectiveness of the SSRIs, but for their safety. Unlike MAOIs and TCAs, the SSRIs are rarely toxic even when taken in overdose. The SSRIs, which includes fluoxetine, sertraline, paroxetine, citalopram, and fluvoxamine, also tend to have fewer adverse effects on weight and serum lipids than the TCAs.

Most patients experience little if any weight gain with the SSRIs. However, with long-term treatment, about 15 percent of patients can have significant increases in weight[6]. The SSRI most likely to cause weight gain is paroxetine, whereas those least likely to increase weight are fluoxetine and citalopram[7]. There are reports that SSRIs can slightly improve insulin sensitivity and glucose tolerance[8], which would tend to improve Metabolic Syndrome. Fluoxetine, in particular, has been found to slightly enhance insulin sensitivity in diabetic patients[9]. Overall, however, the effects of SSRIs on insulin resistance are not remarkable. Paroxetine has been reported to increase serum cholesterol; however, because increase includes both HDL and LDL, the clinical significance is not clear[10]. The new class of antidepressants known as serotonin and norepinephrine reuptake inhibitors (SNRIs) includes venlafaxine and duloxetine. Neither of these medicines has significant effect on weight[11], nor do they affect insulin or glucose metabolism[12]. Slight increases in cholesterol that have occassionaly been seen with these drugs are too small to be of clinical significance[13].

Bupropion is an antidepressant that has been thought to inhibit reuptake of norepinephrine and stimulate release of dopamine[14]. However, its mechanism of action remains unknown. Bupropion has some distinct advantages for certain patients. It rarely causes weight gain, and can often help patients lose weight[15]. It also has few sexual side effects. Of some significance is the fact that bupropion is the same medicine as Zyban, which is used to help people stop smoking. A disproportionately large percentage of patients with chronic depression are smokers, and this habit exacerbates Metabolic

Syndrome. Bupropion has no significant effects on insulin activity[16], nor does it affect serum cholesterol levels[17].

The unique antidepressant mirtazapine increases appetite, and can cause considerable weight gain in some individuals. It is rarely my first choice to treat depressed adults who are otherwise healthy. Young women who are deathly afraid of gaining a few pounds are certainly not good candidates. However, I confess a fondness for this medicine. It is a good antidepressant. It helps anxiety, and it helps people sleep. It is ideal for patients who are depressed, along with having serious medical problems, for often such patients have difficulty with both sleep and appetite. It is also a good choice in treating elderly patients who are depressed, not sleeping, and refusing to eat. It is likely the blocking of histamine-1 and 5-HT2C receptors in the brain by mirtazapine that aids sleep and increases appetite. Although mirtazapine causes weight gain, it does not directly alter insulin sensitivity, glucose metabolism, or cholesterol levels[18].

Trazodone is an old antidepressant that is now used more to treat insomnia than depression. Nonetheless, it has a unique mechanism of action that can be useful as monotherapy or as augmentation for other antidepressants. Like most other sedating antidepressants, it blocks histamine receptors. However, it has far less affinity for the H1 receptor than do the TCAs[19]. Trazodone produces much of its beneficial effect by blocking 5-HT2 receptors. Perhaps, as a consequence of action at those receptors, many patients receiving trazodone in the large doses necessary for its antidepressant action report increases in appetite and weight gain. In one study, trazodone caused an average gain of 1.2 lb after 6 weeks[20]. The bright side is that the drug may decrease serum cholesterol[21].

MOOD STABILIZERS

As with antidepressants, some of the medications used to treat BPAD can cause Metabolic Syndrome. Lithium is a drug that has been remarkably successful in the treatment of BPAD since it was first used in the 1950s. Curiously, lithium mimics some of the effects of insulin in the brain. In muscle tissue lithium can enhance some of insulin's effects[22]. Lithium can sometimes cause weight gain[23]. However, it does not tend to cause or worsen any other aspects of Metabolic Syndrome.

Lamotrigine is a relatively new medication for the treatment of BPAD. It does not cause weight gain, nor does it appear to cause insulin resistance[24]. Topiramate, which is sometimes used without FDA approval for treatment of BPAD, may actually cause weight loss and enhance insulin sensitivity[25]. It is one of the medications that can be used successfully to minimize the weight gain caused by atypical antipsychotics. Carbamazepine has been reported to cause weight gain[26]. However, neither carbamazepine nor the closely related medication oxycarbazepine affect insulin sensitivity or serum levels of triglycerides and cholesterol[27].

Of all the anticonvulsant mood stabilizers, the most adverse effects are seen with valproic acid. Valproic acid can cause weight gain[28], insulin resistance, hypertriglyceridemia, and other components of Metabolic Syndrome[29].

It decreases production of adiponectin, which is an adipocytokine that enhances insulin sensitivity[30]. Valproic acid also increases the likelihood of PCO[31], which is a syndrome of insulin resistance, obesity, infertility, and mood disorder. Slow release forms of valproic acid may be more benign.

In view of the usefulness of anticonvulsant medications in the treatment of BPAD, and ability of the so-called ketogenic diet to treat difficult cases of epileptic seizure disorders, there has been speculation about whether the ketogenic diet might be useful in treating BPAD. As I will discuss in Chapter 9, ketogenic diets can be very useful in the initial steps of losing weight and bringing Metabolic Syndrome under control. Theoretically, it could be used to treat both BPAD and Metabolic Syndrome. Unfortunately, the only report in the literature of placing a patient with BPAD on a ketogenic diet described the treatment as ineffective[32].

Of particular concern in the treatment of BPAD is the use of so-called atypical antipsychotics. This class of medications includes olanzapine, risperidone, quetiapine, ziprasidone, and aripiprazole. These medications were initially developed as anitpsychotics to treat schizophrenia. However, they are also extremely useful in treating BPAD and treatment-resistant Major Depression. Although these medicines can produce remarkable effects, they can also cause serious and even life-threatening side effects if not monitored properly.

ATYPICAL ANTIPSYCHOTICS

For the last 50 years schizophrenia has been treated with medications that block the effects of dopamine in the brain. It has come to be recognized that schizophrenia is not simply a matter of having too much dopamine in the brain. In recent years, a variety of other neurochemical abnormalities have been suspected of contributing to the illness, and many of the current medications reflect this complexity in having extremely complicated pharmacological profiles. Nonetheless, all effective treatments for schizophrenia still include some degree of blockade of dopamine activity in the brain.

The first dopamine antagonists used in treating schizophrenia, such as chlorpromazine, and haloperidol, gave tremendous benefits to many sufferers. Some were able to leave mental hospitals and return home. However, there could be dreadful side effects of these medications. Some patients developed movement disorders. Some were relieved of delusions and hallucinations, but only at doses that sedated them beyond their ability to function in society. Whereas some symptoms of schizophrenia were improved by the medications, little change was seen in others. It was the "positive" symptoms, such as agitation, hallucinations, and delusions, that were best relieved by these drugs. The "negative" symptoms, including lack of motivation, slowness of thought, and social withdrawal, were less improved by the dopamine antagonists then available.

In 1972, a new medicine came into use for the treatment of schizophrenia. This medication, clozapine, was effective in some patients that had received no benefits from the older, "typical" antipsychotics. In some patients, the benefits of clozapine also included relief from negative symptoms. Of striking significance was the fact that clozapine did not produce the disorders of movement that often resulted from treatment with the older antipsychotics.

Unfortunately, clozapine had its own unique set of side effects. The most significant one was the rare, but potentially fatal blood disorder, agranulocytosis. This condition results in loss of white blood cells and subsequent inability to fight off infection. There were a number of deaths from this cause. When clozapine is used, often as a last resort, blood tests are routinely performed to avoid agranulocytosis. Aside from the potential fatal consequences of clozapine, the more frequently noted side effects were weight gain and diabetes.

Scientists avidly pursued the development of medications with the efficacy of clozapine, but without the risk causing agranulocytosis. They produced a cadre of what are commonly called "atypical" antipsychotics. Like clozapine, these medications are useful for treating both positive and negative symptoms of schizophrenia. Although not entirely free of risk, they are much less likely to cause movement disorders than the old, typical antipsychotics. Most importantly, they are safer than clozapine, as they do not cause agranulocytosis. Unfortunately, like clozapine, some of the new atypical antipsychotics stimulate weight gain, and contribute to both Metabolic Syndrome and diabetes.

ATYPICAL ANTIPSYCHOTICS AND WEIGHT GAIN

The older, typical antipsychotics, such as haloperidol, chlorpromazine[33], and thioridazine[34] can cause substantial weight gain in some patients. However, with clozapine and other atypical antipsychotics, weight gain can be dramatic. Almost half of patients taking clozapine have weight increases of more than 10 percent of their initial weight. The next most likely to cause weight gain is olanzapine. Nearly 30 percent of patients taking olanzapine can expect to gain more than 10 percent of their intial weight[35]. There are differences among the atypical antipsychotics in their tendencies to cause such problems. Although quetiapine and risperidone cause weight gain in some patients, they are less likely to do so than olanzapine. Aripiprazole and ziprasidone have little if any affect on weight[36]. I should add a caveat about ziprasidone. With time it has become apparent that the medication is underdosed. It is not yet clear if it will continue to have a benign metabolic profile when higher doses come into greater use.

The "new" antipsychotic, paliperidone, is actually an active metabolite of the old atypical antipsychotic, risperidone. A brief, 6-week trial led authors to conclude that paliperidone did not cause weight gain or "glucose-related" adverse effects[37]. I must confess a certain skepticism at this point. The binding profiles of paliperidone and risperidone are very similar[38], and I have no basis to believe that they would not cause the same weight gain, dyslipidemia and disturbances of energy metabolism. Time will tell.

GENETICS AND ATYPICAL ANTIPSYCHOTIC-INDUCED WEIGHT GAIN

A puzzling aspect of the effects of atypical antipyschotics on weight is that these medications can affect different people in quite different ways. For example, in a recent study 18 percent of subjects taking olanzapine gained more than

20 lb over a year's time. However, in the same study, 11 percent of subjects prescribed olanzapine actually lost more than 10 lb[39].

In my conversation with Gerald Reaven, he expressed the opinion that the effects of olanzapine and other atypicals on weight gain and Metabolic Syndrome are likely to vary in the same way that insulin sensitivity varies in the population. Some patients taking atypical antipsychotics will develop obesity and diabetes, and some will not because of genetic predisposition. Recent studies on the genetics of atypical antipsychotic-induced weight tend to support Reaven's suspicion.

There is evidence that implicates several specific genes as potential mediators of some of the adverse metabolic effects of the atypicals. Among these are genes that code for the serotonin 5-HT2C receptor, leptin, adrenergic alpha2a receptor (ADRa2a), synaptosomal-associated protein 25Da (SNAP25), and the intracellular second messenger system guanine nucleotide binding protein (GNBP)[40].

5-HT2C receptors and leptin are both important in appetite, obesity and Metabolic Syndrome. It is not difficult to see how mutations of these proteins might predispose to antipsychotic-induced weight gain and pathological changes in insulin sensitivity. It is less clear how ADRa2a and SNAP25 might mediate such effects.

There is no obvious relationship between mutations of ADRa2a and schizophrenia[41] or affective disorders. ADRa2a is most closely linked with Attention Deficit Hyperactivity Disorder[42]. However, along with mediating some effects of norepinephrine in the brain, ADRa2a receptors also play a role in the release of insulin from islet cells of the pancreas[43]. To the best of my knowledge, ADRa2a has not been seen as having a role in Metabolic Syndrome.

SNAP25 is involved in release of dopamine into the synaptic cleft, and abnormalities in its activity have been associated with both schizophrenia and BPAD[44]. It seems more than coincidence that SNAP25, like ADRa2a, is also found in islet cells of the pancreas where it helps regulate release of insulin[45]. Again, however, there is no literature describing a role for SNAP25 in Metabolic Syndrome.

MECHANISMS OF ATYPICAL ANTIPSYCHOTIC-INDUCED WEIGHT GAIN

It is not clear how the atypical antipsychotics cause weight gain. The most obvious explanation for how these medications cause weight gain is by acting as antagonists at histamine[46] receptors in the brain. The ability of the atypicals to cause obesity correlates very well with their affinity for histamine type 1 (H1) receptors in brain tissue[47]. Some pure H1 blocking drugs, for example, astemizole, have also been reported to cause weight gain[48]. That the H1 agonist betahistine can reduce some of the weight gain of patients taking olanzapine[49] further suggests that H1 receptors are involved in the weight gain caused by atypical antipsychotics.

Both leptin and insulin decrease the levels of the enzyme AMP kinase in areas of the hypothalamus that control appetite. A new study has shown that the blocking of H1 receptors by clozapine or olanzapine increases the levels of

AMPK in the same areas of hypothalamus[50]. The atypical antipsychotics less active at H1 receptors, including quetiapine, risperidone and ziprasidone, did not cause increases in AMPK. (To *my* eye, the Western Blot assays presented in the published paper suggest that some such changes did occur after treatment with quetiapine and risperidone, although those changes may not have reached statistical significance.) An interesting finding in the study noted above was that the decreases in hypothalamic AMPK produced by leptin and insulin were reversed by treatment with clozapine. This is, de facto, an example of insulin resistance produced by this class of drugs. It leads me to suspect that individuals already exhibiting some form of insulin resistance may be predisposed to the weight gain that can be caused by atypical antipsychotics. An interesting question is that if the effects of clozapine and olanzapine on hypothalamic AMPK levels are so substantial, then why do some patients fail to gain weight or even lose weight while taking these medications?

Antagonism of activity at serotonin type 2C (5-HT2C) receptors in the brain has also been offered as a mechanism of atypical-induced obesity[51]. Mice genetically engineered to not express 5-HT2C receptors exhibit obesity and other components of Metabolic Syndrome[52]. Clozapine and olanzapine have very high affinities for both H1 and 5-HT2C receptors. However, whereas ziprasidone is the atypical antipsychotic with the highest affinity for the 5-HT2C receptor, it is also the atypical antipsychotic least likely to produce weight gain. Furthermore, there is no evidence to suggest differences among the atypicals in their actions at either 5-HT2C or 5-HT2A receptors. They are all antagonists. Thus, questions arise as to how important this receptor could be in mediating the weight gain they cause. The typical antipsychotic molindone has often been reported to cause weight *loss* rather than gain[53]. To the best of my knowledge, this remarkable effect has not been explained. However, it could be due to the fact that molindone has little effect at either 5-HT2[54] or H1[55] receptors in the brain.

It has also been suggested that atypical antipsychotics might cause weight gain by altering the actions of hormones that control hunger and satiety. The data are inconsistent. In some studies, the appetite-stimulating hormone ghrelin is found to increase in patients taking olanzapine[56], whereas in others it decreases or remains unchanged[57]. Serum leptin levels tend to increase with olanzapine[58] and other atypicals[59]. Serum leptin increases as a function of the amount of fat being stored away in adipocytes. Thus, antipsychotic-induced increases in serum leptin could be the *result* rather than the *cause* of increases in body fat. Nonetheless, small but significant increases in serum leptin have been observed during treatment with clozapine and olanzapine before increases in weight[60], which would tend to implicate leptin as a contributing factor. Persistent increases in serum leptin leads to leptin resistance, which can in turn lead to increases in food intake due to disinhibition of hypothalamic appetite centers. Once established, leptin resistance could make it more difficult to lose the weight that has been gained.

Atypical Antipsychotics and Insulin Resistance

It is likely that atypical antipsychotics cause weight gain by blocking activation of central H1 and 5-HT2C receptors, which, in turn, increases appetite.

Weight gain and obesity might then play a role in the development of insulin resistance and Metabolic Syndrome that occurs in patients treated with these medications. However, changes in insulin sensitivity and energy metabolism can sometimes be seen in the absence of weight gain in patients treated with atypical antipsychotics. Thus, these medications might act directly to alter insulin activity.

Measurements of glucose uptake in humans have shown that olanzapine causes insulin resistance[61]. However, it is not clear how this occurs. Supratherapeutic concentrations of olanzapine and other antipsychotics increase insulin resistance by interfering with translocation of GLUT IV glucose transporters to the surface of cells[62]. At therapeutic doses of those medications, this does not occur[63]. Olanzapine has been found to inhibit insulin-induced stimulation of glycogen synthesis in muscle cells[64], and to block the ability of insulin to turn off production of glucose by the liver[65]. Both of those effects would increase serum glucose and require the release of additional insulin to normalize glucose levels. Risperidone[66] and quetiapine[67] can cause insulin resistance; however, they are less likely to do so than olanzapine. Neither ziprasidone[68] nor aripiprazole[69] have significant effects on the body's sensitivity to insulin.

The 5-HT2A receptor antagonist, ketanserin, causes insulin resistance at relatively low doses in human subjects[70]. Because all of the atypical antipsychotics block the 5-HT2A receptor to some degree, it has been speculated that this mechanism might be partially responsible for the insulin resistance these medications can produce. However, ziprasidone, which does not cause insulin resistance, has substantially higher affinity for the 5-HT2A receptor than quetiapine, which can reduce the sensitivity to insulin.

Some atypical antipsychotics may increase resistance to insulin by adversely affecting serum levels of adipocytokines that modulate energy metabolism. Olanzapine decreases adiponectin, an adipocytokine that enhances sensitivity to insulin[71]. At the same time, olanzapine and other atypical antipsychotics increase serum levels of resistin and TNF-alpha[72], both of which increase insulin resistance[73]. The combined effect would be increases in insulin resistance.

The action of insulin is complicated by the fact that when it stimulates the insulin receptor, many secondary biochemical pathways are activated and processed in parallel. Multiple pathways allow the possibility of a medication enhancing some effects of insulin within a cell while inhibiting others. The mood stabilizers, lithium and valproic acid, and the atypical neuroleptics olanzapine, clozapine, quetiapine, risperidone, and ziprasidone all mimic insulin by inhibiting GSK-3 in mouse brain[74].

However, while olanzapine mimics insulin by inhibiting GSK3, it may also antagonize some of insulin's effects, including insulin-induced activation of PI3K. Amisulpride is an atypical antipsychotic used in Europe that does not cause diabetes. Unlike olanzapine, it does not inhibit the activation of PI3K by insulin[75].

Effects of Atypical Antipsychotics on Serum Lipids

Differences among atypical antipsychotics in their effects on insulin activity and weight gain are reflected in adverse changes they produce in HDL, LDL and serum triglyceride levels. Along with hypertension, these abnormalities are the criteria that define Metabolic Syndrome. There is a close relationship between tendency to cause weight gain and subsequent hyperlipidemia. One study compared olanzapine, which tends to cause weight gain, with ziprasidone, which does not. Olanzapine increased total cholesterol, LDL, and triglycerides, whereas Ziprasidone decreased LDL, increased HDL, and had no adverse effects on triglycerides[76]. Triglycerides and weight may also improve after switching from olanzapine to risperidone[77]. In a study comparing a variety of atypical antipsychotics, clozapine and olanzapine produced "marked", and quetiapine "modest", increases in triglycerides. Risperidone had "minimal" effects on triglycerides[78]. Aripiprazole does not appear to affect cholesterol or triglycerides[79].

It should be understood that development of Metabolic Syndrome is a cumulative process. Unlike weight gain that can be seen in patients within weeks after starting atypical antipsychotics, it is likely that the incidence of Metabolic Syndrome grows slowly, but steadily with time.

Atypical Antipsychotics and Diabetes

Insulin resistance, insufficent release of insulin, and hyperglycemia are the hallmarks of Diabetes Type II. This form of diabetes is what occurs in patients with Metabolic Syndrome when their pancreases can no longer secrete the amount of insulin required by the insulin-resistant body to maintain normal serum glucose levels. Diabetes Type II is the concern for patients taking atypical antipsychotics. Atypical antipsychotics rarely cause complete cessation of insulin release from the pancreas, as might be found in Diabetes Type I. Nonetheless, it may occur in some cases.

Almost from their introduction in the 1950s, there have been concerns that antipsychotic drugs can cause or exacerbate Diabetes Type II[80]. Some atypical antipsychotics are particularly prone to causing this form of diabetes, with Clozapine being most likely to do so. Olanzapine is next in the likelihood of producing this diabetes, and it is followed in order of risk by quetiapine, risperidone, and ziprasidone. Some experts conclude that only clozapine and olanzapine cause risk beyond what is seen in untreated patients[81].

Although olanzapine does not change baseline levels of insulin in schizophrenic patients, it reduces the amount of insulin released from the pancreas in response to glucose loading. This effect was not seen after treatment with risperidone[82]. A blunted insulin response to a glucose load would result in hyperglycemia, which is a common side effect of olanzapine[83]. However, there are rare reports of ketoacidosis in individuals recently started on clozapine[84] or olanzapine[85]. Ketoacidosis is most often seen in people with Diabetes Type I, in whom there is little if any insulin being produced by the pancreas. In Diabetes Type II there is usually enough insulin present to prevent this condition. The rare cases of ketoacidosis seen with olanzapine may be due to its decreasing the release of insulin from the pancreas.

There is little evidence that atypical antipsychotics other than olanzapine inhibit the release of insulin. Thus, it is puzzling why quetiapine and risperidone[86] have, on rare occasion, been reported to cause ketoacidosis. Because aripiprazole has not been associated with obesity, Metabolic Syndrome or diabetes, it is quite surprising that there have been at least two reports of ketoacidosis in patients started on this medication[87]. To the best of my knowledge, there have been no reports of ketoacidosis attributed to the use of ziprasidone.

Ketoacidosis does not generally occur in individuals with mere insulin resistance. It is a danger that evolves due to the lack of insulin itself. ADRa2a and SNAP25 have both been implicated as potential genetic factors in antipsychotic-induced weight gain. It is reasonable to assume that both are affected by atypical antipsychotics. It is not likely that ADRa2a- and SNAP25-mediated weight gain could cause ketoacidosis. However, both ADRa2a and SNAP25 are found in the pancreas where they are involved in modulating the release of insulin. That aspect of their activity may be a link between atypical antipsychotics and ketoacidosis.

Another explanation for ketoacidosis could be decreases in insulin release secondary to antimuscarinic effects of some atypical antipsychotics[88]. The parasympathetic nervous system, whose primary neurotransmitter is acetylcholine, stimulates the pancreas to release insulin. Muscarinic type 3 (M3) receptors in the beta cells of the pancreas mediate these effects of acetylcholine. The atypicals with the highest affinity for M3 are also the ones most often reported to cause ketoacidosis, that is, clozapine and olanzapine. Aripirazole, which has been reported to cause ketoacidosis in two cases, has significant antimuscarinic effects, as does quetiapine. The atypicals with lowest affinity for M3 receptors are ziprasidone and risperidone. Ketoacidosis has not been reported to occur with ziprasidone, although it has been seen in a few patients taking risperidone.

Hyperprolactinemia

Prolactin is a hormone released from the anterior pituitary gland. Its primary role is to stimulate breast tissue during development and milk production during lactation. Prolactin also modulates the release of luteinizing hormone (LH) from the pituitary gland. In women, LH stimulates the ovary to produce estrogen and progesterone, and to ovulate. High levels of prolactin inhibit the release of LH from the pituitary, which is partially responsible for the relative lack of fertility in women nursing infants. As many women have learned, it is not a reliable method of birth control. In men, LH is primarily responsible for testosterone production. Men with high levels of serum prolactin tend to have low levels of testosterone and low libido. However, the role of prolactin in human physiology is likely more complex than has generally been suspected. For example, adipocytes have prolactin receptors, and the hormone can modulate adipogenesis, lipogenesis, and lipolysis in those cells[89]. Prolactin can also affect receptors for insulin-like growth hormone and other growth factors in the liver[90]. In patients that are neither psychotic nor taking antipsychotic medication, hyperprolactinemia is associated with many components of Metabolic Syndrome. Prolactin reduces insulin sensitivity; increases serum insulin, glucose, C-reactive protein, and homocysteine levels[91]; and lowers serum HDL[92].

A major component of the control of serum levels of the hormone is tonic inhibition of its release by dopamine. The dopamine is supplied by the tuberoinfundibular component of the brain's network of dopaminergic neurons. The inhibitory effects of dopamine on prolactin release are mediated by the D2 receptor. A common treatment of hyperprolactinemia, often secondary to prolactin-releasing microadenomas in the pituitary gland, is oral administration of D2 agonists such as bromocryptine. It is worth noting that in otherwise normal obese women with hyperprolactinemia, signs of Metabolic Syndrome are seen in proportion to the degree that they exhibit visceral obesity. Treatment of those women with the D2 agonist bromocryptine both decreases their serum prolactin levels and helps reverse some aspects of Metabolic syndrome including hyperinsulinemia, hyperglycemia, and hypertension[93].

Blockade of the D2 receptor is a component of the action of every antipsychotic medication. The older, typical antipsychotics tend to act primarily at D2 receptors, and thus they produce significant degrees of hyperprolactinemima. The newer, atypical antipsychotics are also D2 antagonists but are less dependent on this aspect of their pharmacological profile in producing the desired antipsychotic effect. Thus, they tend to produce their therapeutic effects with less likelihood of hyperprolactinemia. It has even been suggested that decreases in the probability of hyperprolactinemia is another feature of "atypicality" in antipsychotic medications[94]. There are several exceptions, and they illuminate the present discussion.

It would be reasonable to suspect that hyperprolactinemia could play a role in antipsychotic-induced weight gain and Metabolic Syndrome. However, there is no relationship between the ability of an antipsychotic to increase prolactin, and the degree to which it causes weight gain and other aspects of Metabolic Syndrome. Clozapine[95] and olanzapine, the antipsychotics most likely to cause weight gain and Metabolic Syndrome, have little effect on serum prolactin. Amisulpride, an atypical antipsychotic used in Europe, causes substantial hyperprolactinemia but has few metabolic effects[96]. Risperidone and quetiapine have about the same propensity to cause weight gain and Metabolic Syndrome, but of the two, only risperidone dramatically increases prolactin levels[97]. The typical antipsychotic molindone raises serum prolactin[98], yet it can cause weight loss in some patients. The data clearly indicate that while hyperprolactinemia could contribute to the adverse metabolic effects of certain antipsychotics, it is not a major factor in causing Metabolic Syndrome in this class of medications.

GUIDELINES FOR MONITORING PATIENTS ON ATYPICAL ANTIPSYCHOTICS

An expert panel that arose out of collaborations between the American Psychiatric Association and the American Diabetes Association has constructed guidelines to help avoid adverse metabolic effects and Metabolic Syndrome when prescribing atypical antipsychotics[99]. The focus of this panel was the chronic use of antipsychotics in the treatment of schizophrenia. However, the guidelines are useful for any prolonged use of this class of medications, including their use in treating BPAD or as augmentation strategies to control treatment-resistant depression.

Prior to initiating treatment with atypical antipsychotics, one should obtain baseline measurements of weight, BMI, waist circumference, blood pressure, fasting lipids, and fasting serum glucose. There should also be a review of the patient's personal and family medical history to ascertain the degree to which they might be predisposed to Metabolic Syndrome. After starting the atypical antipsychotic, measurements of weight and blood pressure should be obtained on a monthly basis, and fasting lipid and serum glucose levels should be evaluated every 3 months. The final recommendation is to evaluate waist circumference on an annual basis.

The thresholds of concern for each of those parameters while prescribing an atypical antipsychotic are simply those defined in the diagnostic criteria for Metabolic Syndrome. Although weight, *per se*, is not a parameter used in defining Metabolic Syndrome, it is suggested that a 5 percent increase in weight should trigger concern that an adverse effect on weight has emerged. The consensus is that any adverse changes in the baseline parameters should first be addressed by changes in lifestyle. These lifestyle changes involve healthier approaches to diet, exercise, sleep, and stress control, as well as the elimination of bad habits. There is preliminary evidence that simply educating patients about more healthy diet and other lifestyle choices can significantly reduce the weight gain that often accompanies treatment with the atypical antipsychotics[100]. I discuss specific changes that can be made in Chapter 11 of this book.

If the adverse changes cannot be brought under control after 3 months of trying, it is recommended that other means be applied to solve the problem. This might include switching the atypical antipsychotic to one with a more benign metabolic profile. If such a switch puts the patient in jeopardy of decompensation, then it might be necessary to maintain the atypical antipsychotic but add medication to specifically treat weight gain, hyperglycemia, or hyperlipidemia.

ADD-ON TREATMENTS TO COUNTER WEIGHT GAIN AND METABOLIC SYNDROME

Several different medications have been added on to atypical antipsychotics to counter their tendency to cause weight gain. The one that has most often been used for this purpose is topiramate. Topiramate can minimize or even reverse weight gain in schizophrenic patients induced by olanzapine[101] or risperidone. An extra benefit for some patients is that topiramate may help improve both positive and negative symptoms of treatment-resistant schizophrenics when it is added to their existing antipsychotic regimen[102]. Topiramate is also helpful in reducing olanzapine-induced weight gain in patients being treated for BPAD[103]. Amantadine, with at least one exception[104], has also been found to help decrease weight gain in patients treated with olanzapine[105]. This affect of amantadine was not accompanied by any decreases in the antipsychotic efficacy of olanzapine. The mechanisms by which topiramate and amantadine reverse weight gain induced by atypical antipsychotics are not clear.

Betahistine, the H1 agonist, can also slightly decrease olanzapine-induced weight gain. It has not been as effective as topiramate or amantadine. I should note that betahistine has been found to increase the likelihood of gastric ulcers, and its use needs to be monitored. The H2 receptor blockers, nizatidine[106] and famotidine[107], appear to have little if any effect on atypical antipsychotic-induced weight gain.

The oral hypoglycemic agent, metformin, was found to both decrease weight gain and enhance insulin-sensitivity in children taking several different types of atypical antipsychotics[108]. Unfortunately, it had neither such effect in adult schizophrenics prescribed olanzapine[109]. The only beneficial effect in those adult patients was normalization of serum glucose. At least one study has found that the monoamine reuptake inhibitor sibutramine, which is marketed as a weight loss drug, can reduce weight gained by patients taking olanzapine[110]. Moreover, it may do so without exacerbating psychotic symptoms.

I am not aware of any reports of the opiate antagonist, naltrexone, being useful in treating weight gain in schizophrenia. However, there are reports of naltrexone helping to reverse weight gain in patients treated with TCAs and lithium[111]. There are scattered, although weakly substantiated, reports that naltrexone can be helpful as an augmentation strategy in schizophrenia[112]. In any case, it does not appear to cause harm in such patients, and has been used for substance abuse problems in schizophrenics also prescribed antipsychotics[113]. Thus, naltrexone is an unevaluated, but potentially useful medication to help contain weight gain in patients being treated with atypical antipsychotics.

Another promising, but as of yet untested, treatment for antipsychotic-induced obesity and Metabolic Syndrome is the new cannabinoid receptor (CB1) antagonist, rimonabant. Because of the peculiar effects that cannabinoid agonists can have on thought and perception, it has long been suspected that CB receptors might play a role in the etiology of schizophrenia and other psychiatric illnesses. There are no indications that atypical antipsychotics act on CB1 receptors [114], nor have CB1 antagonists lived up to initial expectations of having antipsychotic properties[115]. Nonetheless, rimonabant and other CB1 antagonists have potent ability to reduce appetite and reverse some of the cardinal components of Metabolic Syndrome, including abdominal obesity[116], insulin resistance, atherogenic dyslipidemia, and inflammation[117]. Cannabinoid agonists may exacerbate some symptoms of schizophrenia, which would lead one to suspect that adding a CB1 antagonist would not hurt and could possibly even help some patients with schizophrenia. On the other hand, some symptoms of BPAD may be improved by stimulation of cannabinoid receptors[118]. It will be interesting to see if rimonabant will be of equal usefulness in controlling antipsychotic-induced Metabolic Syndrome in schizophrenia and BPAD.

Although there have been no formal studies on the subject, a case report suggests that standard methods of treatment of diabetes, such as the "glitazones" can control hyperglycemia in patients treated with olanzapine[119]. Animal studies have further suggested that glitazones and other PPAR activators can help reverse effects of atypical antipsychotics, such as hyperlipidemia, hyperglycemia, and hyperinsulinemia[120]. Certainly, more study is needed. Nonetheless, I am unaware of ill effects or contraindications for using standard hypoglycemic agents and PPAR activators to reduce adverse metabolic effects of atypical antipsychotics. Because of possible adverse interactions involving competition for P450 enzymes, some care may need to be taken in coadministering statins with risperidone[121] or quetiapine[122] to reduce hyperlipidemia.

DIFFICULT CHOICES MUST BE MADE

In view of the catastrophic effects of schizophrenia, it is essential that patients be treated with antipsychotic medications. Psychiatrists are in a dilemma in deciding which antipsychotic to use. The older typical antipsychotics are somewhat less likely to cause diabetes and weight gain, but they cause more severe movement problems. The newer atypical antipsychotics are less likely to cause movement disorders, and may be more effective for some symptoms of schizophrenia, such as the negative symptoms. However, with the exceptions of ziprasidone and aripiprazole, atypical antipsychotics exacerbate Metabolic Syndrome.

The recent CATIE report[123], in which the atypical antipsychotics were compared with the old, typical antipsychotic perphenazine, has done little to resolve this dilemma. Although there were surprisingly few differences between perphenazine and the newer medications, olanzapine was found to be somewhat more effective than the other antipsychotics in the study. Unfortunately, as is generally found to be the case, olanzapine also caused the most weight gain and adverse metabolic changes.

If perphenazine, ziprasidone and aripiprazole worked well in every patient, the problem would be resolved. However, different patients respond differently to the various types of antipsychotics. Moreover, some patients respond best to the atypical neuroleptics that are most likely to cause obesity, Metabolic Syndrome and diabetes. The fact remains that schizophrenia must be treated and difficult choices must be made. No drugs are without risk, and the benefits and risks must always be carefully weighed.

REFERENCES

1. Lazarou, J., Pomeranz, B.H., and Corey, P.N., Incidence of adverse drug reactions in hospitalized patients: a meta-analysis of prospective studies. *JAMA* 1998; 279:1200–1205.
2. Cantu, T.G. and Korek, J.S., Monoamine oxidase inhibition and weight gain. *Drug Intell. Clin. Pharm.* 1988; 22:755–759.
3. Cooper, A.J. and Ashcroft, G., Potentiation of insulin hypoglycemia by M.A.O.I. antidepressants drugs. *Lancet* 1966; 1:407–409.
4. Frampton, J.E. and Plosker, G.L., Selegiline transdermal system: in the treatment of major depressive disorder. *Drugs* 2007; 67:257–265.
5. Garland, E.J., Remick, R.A., and Zis, A.P., Weight gain with antidepressants and lithium. *J. Clin. Psychopharmacol.* 1988; 8:323–330.
6. Maina, G. et al., Weight gain during long-term treatment of obsessive-compulsive disorder: a prospective comparison between serotonin reuptake inhibitors. *J. Clin. Psychiatry* 2004; 65:1365–1371.
7. Raeder, M.B. et al., Obesity, dyslipidemia, and diabetes with selective serotonin reuptake inhibitors: the Hordaland Health Study. *J. Clin. Psychiatry.* 2006; 67:1974–1982.
8. Araya, V. et al., The effect of fluoxetine on insulin resistance in non-diabetic obese patients. *Rev. Med. Chil.* 1995; 123:943–947.

9. Maheux P, et al., Fluoxetine improves insulin sensitivity in obese patients with non-insulin-dependent diabetes mellitus independently of weight loss. *Int. J. Obes. Relat. Metab. Disord.* 1997; 21:97–102.

10. Kim, E.J., and Yu, B.H., Increased cholesterol levels after paroxetinetreatment in patients with panic disorder. *J. Clin. Psychopharmacol.* 2005; 25:597–599.

11. Kraus, T. et al., Body weight, the tumor necrosis factor system, and leptin production during treatment with mirtazapine or venlafaxine. *Pharmacopsychiatry* 2002; 35:220–225.

12. McIntyre, R.S. et al., The effect of antidepressants on glucose homeostasis and insulin sensitivity: synthesis and mechanisms. *Expert Opin. Drug Saf.* 2006; 5:157–168.

13. McIntyre, R.S. et al., The effect of antidepressants on lipid homeostasis: a cardiac safety concern? *Expert Opin. Drug Saf.* 2006; 5:523–537.

14. Foley, K.F., DeSanty, K.P., and Kast, R.E., Bupropion: pharmacology and therapeutic applications. *Expert Rev. Neurother.* 2006; 6:1249–1265.

15. Gadde, K.M. et al., Bupropion for weight loss: an investigation of efficacy and tolerability in overweight and obese women. *Obes. Res.* 2001; 9:544–551.

16. Othmer, E. et al., Long-term efficacy and safety of bupropion. *J. Clin. Psychiatry* 1983; 44:153–156.

17. Botella-Carretero, J.I. et al., Weight gain and cardiovascular risk factors during smoking cessation with bupropion or nicotine. *Horm. Metab. Res.* 2004; 36:178–182.

18. Laimer, M. et al., Effect of mirtazapine treatment on body composition and metabolism. *J. Clin. Psychiatry* 2006; 67:421–424.

19. Owens, M.J. et al., Neurotransmitter receptor and transporter binding profile of antidepressants and their metabolites. *J. Pharmacol. Exp. Ther.* 1997; 283:1305–1322.

20. Weisler, R.H. et al., Comparison of bupropion and trazodone for the treatment of major depression. *J. Clin. Psychopharmacol.* 1994; 14:170–179.

21. Perry, P.J. et al., A report of trazodone-associated laboratory abnormalities. *Ther. Drug Monit.* 1990; 12:517–519.

22. Tabata, I. et al., Lithium increases susceptibility of muscle glucose transport to stimulation by various agents. *Diabetes* 1994; 43:903–907.

23. Bendz, H. and Aurell, M., [Adverse effects of lithium treatment and safety routines] *Lakartidningen* 2004; 101:1902–1908.

24. Isojarvi, J.I. et al., Valproate, lamotrigine, and insulin-mediated risks in women with epilepsy. *Ann. Neurol.* 1998; 43:446–451.

25. Wilding, J. et al., A randomized double-blind placebo-controlled study of the long-term efficacy and safety of topiramate in the treatment of obese subjects. *Int. J. Obes. Relat. Metab. Disord.* 2004; 28:1399–1410.

26. Jallon, P. and Picard, F., Bodyweight gain and anticonvulsants: a comparative review. *Drug Saf.* 2001; 24:969–978.

27. Pylvanen, V. et al., Fasting serum insulin and lipid levels in men with epilepsy. *Neurology* 2003; 60:571–574.

28. Malhi, G.S., Mitchell, P.B., and Caterson, I., 'Why getting fat, Doc?' Weight gain and psychotropic medications. *Aust. N.Z J. Psychiatry* 2001; 35:315–321.

29. Pylvanen, V. et al., Insulin-related metabolic changes during treatment with valproate in patients with epilepsy. *Epilepsy Behav.* 2006; 8:643–648.

30. Qiao, L., Schaack, J., and Shao, J., Suppression of adiponectin gene expression by histone deacetylase inhibitor valproic acid. *Endocrinology* 2006; 147:865–874.

31. Isojarvi, J.I., et al., Altered ovarian function and cardiovascular risk factors in valproate-treated women. *Am. J. Med.* 2001; 111:290–296.

32. Yaroslavsky, Y., Stahl, Z., and Belmaker, R.H., Ketogenic diet in bipolar illness. *Bipolar Disord.* 2002; 4:75.

33. Zipursky, R.B. et al., Course and predictors of weight gain in people with first-episode psychosis treated with olanzapine or haloperidol. *Br. J. Psychiatry* 2005; 187:537–543.

34. Allison, D.B. et al., Antipsychotic-induced weight gain: a comprehensive research synthesis. *Am. J. Psychiatry* 1999; 156:1686–1696.

35. Lieberman, J.A., Metabolic changes associated with antipsychotic use. *Prim. Care Companion J. Clin. Psychiatry* 2004: 6(Suppl. 2):8–13.

36. Newcomer, J.W., Second-generation (atypical) antipsychotics and metabolic effects: a comprehensive literature review. *CNS Drugs* 2005; 19(Suppl. 1):1–93.

37. Kane, J. et al., Treatment of schizophrenia with paliperidone extended-release tablets: a 6-week placebo-controlled trial. *Schizophr. Res.* 2007; 90:147–161.

38. Leysen, J.E. et al., Risperidone: a novel antipsychotic with balanced serotonin-dopamine antagonism, receptor occupancy profile, and pharmacologic activity. *J. Clin. Psychiatry* 1994; 55(Supp. l):5–12.

39. Lambert, M.T. et al., New-onset type-2 diabetes associated with atypical antipsychotic medications. *Prog. Neuropsychopharmacol. Biol. Psychiatry* 2006; 30:919–923.

40. Muller, D.J. and Kennedy, J.L., Genetics of antipsychotic treatment emergent weight gain in schizophrenia. *Pharmacogenomics* 2006; 7:863–887.

41. Clark, D.A. et al., No association between ADRA2A polymorphisms and schizophrenia. *Am. J. Med. Genet. B Neuropsychiatr. Genet.* 2006; 144B:341–343.

42. Deupree, J.D. et al., Possible involvement of alpha-2A adrenergic receptors in attention deficit hyperactivity disorder: radioligand binding and polymorphism studies. *Am. J. Med. Genet. B Neuropsychiatr. Genet.* 2006; 141:877–884.

43. Clement, L. et al., Pancreatic beta-cell alpha2A adrenoceptor and phospholipid changes in hyperlipidemic rats. *Lipids* 2002; 37:501–506.

44. Kumamoto, N. et al., Hyperactivation of midbrain dopaminergic system in schizophrenia could be attributed to the down-regulation of dysbindin. *Biochem. Biophys. Res. Commun.* 2006; 345:904–909.

45. Chan, C.B. et al., Beta-cell hypertrophy in fa/fa rats is associated with basal glucose hypersensitivity and reduced SNARE protein expression. *Diabetes* 1999; 48:997–1005.

46. Wetterling, T., Bodyweight gain with atypical antipsychotics. A comparative review. *Drug Saf.* 2001; 24:59–73.

47. Richelson, E. and Souder, T., Binding of antipsychotic drugs to human brain receptors focus on newer generation compounds. *Life Sci.* 2000; 68:29–39.

48. Krstenansky, P.M. and Cluxton, R.J. Jr., Astemizole: a long-acting, nonsedating antihistamine. *Drug Intell. Clin. Pharm.* 1987; 21:947–953.

49. Poyurovsky, M. et al., The effect of betahistine, a histamine H1 receptor agonist/H3 antagonist, on olanzapine-induced weight gain in first-episode schizophrenia patients. *Int. Clin. Psychopharmacol.* 2005; 20:101–103.

50. Kim, S.F. et al., Antipsychotic drug-induced weight gain mediated by histamine H1 receptor-linked activation of hypothalamic AMP-kinase. *PNAS* 2007; 104:3456–3459.

51. Pooley, E.C. et al., A 5-HT2C receptor promoter polymorphism (HTR2C - 759C/T) is associated with obesity in women, and with resistance to weight loss in heterozygotes. *Am. J. Med. Genet. B Neuropsychiatr. Genet.* 2004; 126:124–127.

52. Tecott, L.H. et al., Eating disorder and epilepsy in mice lacking 5-HT2c serotonin receptors. *Nature* 1995; 374:542–546.

53. Stanton, J.M., Weight gain associated with neuroleptic medication: a review. *Schizophr. Bull.* 1995; 21:463–472.

54. Wander, T.J. et al., Antagonism by neuroleptics of serotonin 5-HT1A and 5-HT2 receptors of normal human brain in vitro. *Eur. J. Pharmacol.* 1987; 143:279–282.

55. Richelson, E. and Nelson, A., Antagonism by neuroleptics of neurotransmitter receptors of normal human brain in vitro. *Eur. J. Pharmacol.* 1984; 103:197–204.

56. Murashita, M. et al., Olanzapine increases plasma ghrelin level in patients with schizophrenia. *Psychoneuroendocrinology* 2005; 30:106–110.

57. Togo, T. et al., Serum ghrelin concentrations in patients receiving olanzapine or risperidone. *Psychopharmacology (Berl.)* 2004; 172:230–232.

58. Hosojima, H. et al., Early effects of olanzapine on serum levels of ghrelin, adiponectin and leptin in patients with schizophrenia. *J. Psychopharmacol.* 2006; 20:75–79.

59. Haupt, D.W. et al., Plasma leptin and adiposity during antipsychotic treatment of schizophrenia. *Neuropsychopharmacology* 2005; 30:184–191.

60. Kraus, T. et al., Body weight and leptin plasma levels during treatment with antipsychotic drugs. *Am. J. Psychiatry* 1999; 156:312–314.

61. Henderson, D.C. et al., Glucose metabolism in patients with schizophrenia treated with olanzapine or quetiapine: a frequently sampled intravenous glucose tolerance test and minimal model analysis. *J. Clin. Psychiatry* 2006; 67:789–797.

62. Dwyer, D.S. et al., Antipsychotic drugs affect glucose uptake and the expression of glucose transporters in PC12 cells. *Prog. Neuropsychopharmacol. Biol. Psychiatry* 1999; 23:69–80.

63. Robinson, K.A. et al., At therapeutic concentrations, olanzapine does not affect basal or insulin-stimulated glucose transport in 3T3-L1 adipocytes. *Prog. Neuropsychopharmacol. Biol. Psychiatry* 2006; 30:93–98.

64. Engl, J. et al., Olanzapine impairs glycogen synthesis and insulin signaling in L6 skeletal muscle cells. *Mol. Psychiatry* 2005; 10:1089–1096.

65. Houseknecht, K.L. et al., Acute effects of atypical antipsychotics on whole-body insulin resistance in rats: implications for adverse metabolic effects. *Neuropsychopharmacology* 2007; 32:289–297.

66. Wu, R.R. et al., Effects of typical and atypical antipsychotics on glucose-insulin homeostasis and lipid metabolism in first-episode schizophrenia. *Psychopharmacology (Berl.)* 2006; 186:572–578.

67. Henderson, D.C. et al., Glucose metabolism in patients with schizophrenia treated with olanzapine or quetiapine: a frequently sampled intravenous glucose tolerance test and minimal model analysis. *J. Clin. Psychiatry* 2006; 67:789–797.

68. Weiden, P.J. et al., Best clinical practice with ziprasidone: update after one year of experience. *J. Psychiatr. Pract.* 2002; 8:81–97.

69. Kerpel-Fronius, S. and Lóránt, M., [A new atypical antipsychotic with partial dopamine agonist effect (aripiprazole)]. *Neuropsychopharmacol. Hung.* 2004; 6:177–184.

70. Gilles, M. et al., Antagonism of the serotonin (5-HT)-2 receptor and insulin sensitivity: implications for atypical antipsychotics. *Psychosom. Med.* 2005; 67:748–751.

71. Richards, A.A. et al., Olanzapine treatment is associated with reduced high molecular weight adiponectin in serum: a potential mechanism for olanzapine-induced insulin resistance in patients with schizophrenia. *J. Clin. Psychopharmacol.* 2006; 26:232–237.

72. Birkas Kovats, D. et al., Possible connection between ghrelin, resitin and TNF-alpha levels and the metabolic syndrome caused by atypical antipsychotics. *Neuropsychopharmacol. Hung.* 2005; 7:132–139.

73. Meier, U. and Gressner, A.M., Endocrine regulation of energy metabolism: review of pathobiochemical and clinical chemical aspects of leptin, ghrelin, adiponectin, and resistin. *Clin. Chem.* 2004; 50:1511–1525.

74. Li, X. et al., Regulation of mouse brain glycogen synthase kinase-3 by atypical antipsychotics. *Int. J. Neuropsychopharmacol.* 2006; 4:1–13.

75. Engl, J. et al., Olanzapine impairs glycogen synthesis and insulin signaling in L6 skeletal muscle cells. *Mol. Psychiatry* 2005; 10:1089–1096.

76. Brown, R.R. and Estoup, M.W., Comparison of the metabolic effects observed in patients treated with ziprasidone versus olanzapine. *Int. Clin. Psychopharmacol.* 2005; 20:105–112.

77. Su, K.P., Wu, P.L., and Pariante, C.M., A crossover study on lipid and weight changes associated with olanzapine and risperidone. *Psychopharmacology (Berl.)* 2005; 183:383–386.

78. Atmaca, M. et al., Serum leptin and triglyceride levels in patients on treatment with atypical antipsychotics. *J. Clin. Psychiatry* 2003; 64:598–604.

79. McQuade, R.D. et al., A comparison of weight change during treatment with olanzapine or aripiprazole: results from a randomized, double-blind study. *J. Clin. Psychiatry* 2004; 65(Suppl. 18):47–56.

80. Dunlop, B.W. et al., Disturbed glucose metabolism among patients taking olanzapine and typical antipsychotics. *Psychopharmacol. Bull.* 2003; 37:99–117.

81. Gianfrancesco, F. et al., Assessment of antipsychotic-related risk of diabetes mellitus in a Medicaid psychosis population: sensitivity to study design. *Am. J. Health Syst. Pharm.* 2006; 63:431–441.

82. Chiu, C.C. et al., The early effect of olanzapine and risperidone on insulin secretion in atypical-naive schizophrenic patients. *J. Clin. Psychopharmacol.* 2006; 26:504–507.

83. Seemuller, F. et al., The safety and tolerability of atypical antipsychotics in bipolar disorder. *Expert Opin. Drug Saf.* 2005; 4:849–868.

84. Jin, H., Meyer, J.M., and Jeste, D.V., Phenomenology of and risk factors for new-onset diabetes mellitus and diabetic ketoacidosis associated with atypical antipsychotics: an analysis of 45 published cases. *Ann. Clin. Psychiatry* 2002; 14:59–64.

85. Howes, O.D. and Rifkin, L., Diabetic keto-acidotic (DKA) coma following olanzapine initiation in a previously euglycaemic woman and successful continued therapy with olanzapine. *J. Psychopharmacol.* 2004; 18:435–437.

86. Dibben, C.R. et al., Diabetes associated with atypical antipsychotic treatment may be severe but reversible: case report. *Int. J. Psychiatry Med.* 2005; 35:307–311.

87. Reddymasu, S. et al., Elevated lipase and diabetic ketoacidosis associated with aripiprazole. *JOP* 2006; 7:303–305.

88. Robinson, D.S., Insulin secretion and psychotropic drugs. *Prim. Psychiatry* 2006; 13:26–27.

89. Fleenor, D., Arumugam, R., and Freemark, M., Growth hormone and prolactin receptors in adipogenesis: STAT-5 activation, suppressors of cytokine signaling, and regulation of insulin-like growth factor I. *Horm. Res.* 2006; 66:101–110.

90. El Khattabi, I., Remacle, C., and Reusens, B., The regulation of IGFs and IGFBPs by prolactin in primary culture of fetal rat hepatocytes is influenced by maternal malnutrition. *Am. J. Physiol. Endocrinol. Metab.* 2006; 291:835–842.

91. Yavuz, D. et al., Endothelial function, insulin sensitivity and inflammatory markers in hyperprolactinemic pre-menopausal women. *Eur. J. Endocrinol.* 2003; 149:187–193.

92. Heshmati, H.M., Turpin, G., and de Gennes, J.L., Chronic hyperprolactinemia and plasma lipids in women. *J. Mol. Med.* 1987; 65:1432–1440.

93. Kok, P. et al., Activation of dopamine D2 receptors simultaneously ameliorates various metabolic features of obese women. *Am. J. Physiol. Endocrinol. Metab.* 2006; 291:1038–1043.

94. Hamner, M., The effects of atypical antipsychotics on serum prolactin levels. *Ann. Clin. Psychiatry* 2002; 14:163–173.

95. de Leon, J. et al., Possible individual and gender differences in the small increases in plasma prolactin levels seen during clozapine treatment. *Eur. Arch. Psychiatry Clin. Neurosci.* 2004; 254:318–325.

96. Peuskens, J. et al., Metabolic control in patients with schizophrenia treated with amisulpride or olanzapine. *Int. Clin. Psychopharmacol.* 2007; 22:145–152.

97. Goodnick, P.J., Rodriguez, L., and Santana, O., Antipsychotics: impact on prolactin levels. *Expert Opin. Pharmacother.* 2002; 3:1381–1391.

98. Pandurangi, A.K. et al., Relation of serum molindone levels to serum prolactin levels and antipsychotic response. *J. Clin. Psychiatry* 1989; 50:379–381.

99. American Diabetes Association, American Psychiatric Association, American Association of Clinical Endocrinologists, North American Association for the Study of Obesity. Consensus development conference on antipsychotic drugs and obesity and diabetes. *J. Clin. Psychiatry* 2004; 65:1335–1342.

100. Littrell, K.H. et al., The effects of an educational intervention on antipsychotic-induced weight gain. *J. Nurs. Scholarsh.* 2003; 35:237–241.

101. Nickel, M.K. et al., Influence of topiramate on olanzapine-related adiposity in women: a random, double, placebo-controlled study. *J. Clin. Psychopharmacol.* 2005; 25:211–217.

102. Tiihonen, J. et al., Topiramate add-on in treatment-resistant schizophrenia: a randomized, double-blind, placebo-controlled, crossover trial. *J. Clin. Psychiatry* 2005; 66:1012–1015.

103. Vieta, E. et al., Effects on weight and outcome of long-term olanzapine-topiramate combination treatment in bipolar disorder. *J. Clin. Psychopharmacol.* 2004; 24:374–378.

104. Bahk, W.M. et al., Open label study of the effect of amantadine on weight gain induced by olanzapine. *Psychiatry Clin. Neurosci.* 2004; 58:163–167.

105. Graham, K.A. et al., Double-blind, placebo-controlled investigation of amantadine for weight loss in subjects who gained weight with olanzapine. *Am. J. Psychiatry* 2005; 162:1744–1746.

106. Atmaca, M. et al., Nizatidine for the treatment of patients with quetiapine-induced weight gain. *Hum. Psychopharmacol.* 2004; 19:37–40.

107. Poyurovsky, M. et al., The effect of famotidine addition on olanzapine-induced weight gain in first-episode schizophrenia patients: a double-blind placebo-controlled pilot study. *Eur. Neuropsychopharmacol.* 2004; 14:332–336.

108. Klein, D.J. et al., A randomized, double-blind, placebo-controlled trial of metformin treatment of weight gain associated with initiation of atypical antipsychotic therapy in children and adolescents. *Am. J. Psychiatry* 2006; 163:2072–2079.

109. Baptista, T. et al., Metformin for prevention of weight gain and insulin resistance with olanzapine: a double-blind placebo-controlled trial. *Can. J. Psychiatry* 2006; 51:192–196.

110. Henderson, D.C. et al., A double-blind, placebo-controlled trial of sibutramine for olanzapine-associated weight gain. *Am. J. Psychiatry* 2005; 162:954–962.

111. Zimmermann, U. et al., Effect of naltrexone on weight gain and food craving induced by tricyclic antidepressants and lithium: an open study. *Biol. Psychiatry* 1997; 41:747–749.
112. Marchesi, G.F. et al., The therapeutic role of naltrexone in negative symptom schizophrenia. *Prog. Neuropsychopharmacol. Biol. Psychiatry* 1995; 19:1239–1249.
113. Green, A.I., Pharmacotherapy for schizophrenia and co-occurring substance use disorders. *Neurotox. Res.* 2007; 11:33–40.
114. Theisen, F.M. et al., No evidence for binding of clozapine, olanzapine and/or haloperidol to selected receptors involved in body weight regulation. *Pharmacogenomics J.* September 19, 2006 [Epub ahead of print]
115. Meltzer, H.Y. et al., Placebo-controlled evaluation of four novel compounds for the treatment of schizophrenia and schizoaffective disorder. *Am. J. Psychiatry* 2004; 161:975–984.
116. Patel, P.N. and Pathak, R., Rimonabant: a novel selective cannabinoid-1 receptor antagonist for treatment of obesity. *Am. J. Health Syst. Pharm.* 2007; 64:481–489.
117. Duffy, D. and Rader, D., Endocannabinoid antagonism: blocking the excess in the treatment of high-risk abdominal obesity. *Trends Cardiovasc. Med.* 2007; 17:35–43.
118. Ashton, C.H. et al., Cannabinoids in bipolar affective disorder: a review and discussion of their therapeutic potential. *J. Psychopharmacol.* 2005; 19:293–300.
119. Baptista, T. et al., Metformin for prevention of weight gain and insulin resistance with olanzapine: a double-blind placebo-controlled trial. *Can. J. Psychiatry* 2006; 51:192–196.
120. Arulmozhi, D.K., Dwyer, D.S., and Bodhankar, S.L., Antipsychotic induced metabolic abnormalities: an interaction study with various PPAR modulators in mice. *Life Sci.* 2006; 79:1865–1872.
121. Webber, M.A. et al., Rhabdomyolysis and compartment syndrome with coadministration of risperidone and simvastatin. *J. Psychopharmacol.* 2004; 18:432–434.
122. Furst, B.A. et al., Possible association of QTc interval prolongation with co-administration of quetiapine and lovastatin. *Biol. Psychiatry* 2002; 51:264–265.
123. Lieberman, J.A., Effectiveness of antipsychotic drugs in patients with chronic schizophrenia. *N. Engl. J. Med.* 2005; 353:1209–1223.

6

DEPRESSION, METABOLIC SYNDROME, AND HEART DISEASE

Poets and kindly grandmothers have long known that it is possible to die from a broken heart. However, this possibility came under scientific scrutiny after a flurry of papers began to provide evidence of a physiological connection between depression and heart disease. In the 1970s, Danish studies showed that Major Depression predisposed patients to develop cardiovascular disease[1]. A paper published in 1988 by Dr. Robert Carney of the Washington University School of Medicine concluded that Major Depression was as great a risk for heart attack as cigarette smoking in people already known to have coronary artery disease[2]. It has recently been shown that simply having a pessimistic outlook on life increases the risk of death from heart disease[3]. Thirty years of medical research has clearly established that individuals with Major Depression are more likely than those without depression to develop and die from cardiovascular disease. When patients have already had a myocardial infarction, a diagnosis of Major Depression makes it more likely that they will have a second one[4]. The question has now become, "What is it about Major Depression that makes heart disease worse?"

As many as 27 percent of patients with heart disease also have Major Depression[5]. This is substantially higher than the 5 percent of adults that have been estimated to have Major Depression at any one time in the United States[6]. Of course, having a severe illness would be expected to cause depressed mood, and reports of depression among patients with heart disease are similar to those reported among cancer patients. However, a sobering fact is that adults with depression and no heart disease are two to three times as likely to later be diagnosed with heart disease than are people without depression. This risk was found to be independent of any other known risks for the disease. In patients already diagnosed with heart disease, or who have already had a heart attack, the additional diagnosis of Major Depression increases the risk of dying within 2 years by nearly three-fold[7].

As with the relationship between depression and Metabolic Syndrome, it was initially assumed that depression exacerbates heart disease because depressed people are less able and willing to follow their doctor's orders. They do not eat right or exercise enough. They smoke. They fail to take their medication properly and miss doctor's appointments. However, even when these problems were taken into account, it was found that something about depression itself made the prognosis worse in patients with heart disease. It was then that the connections between depression, stress, inflammation, and other contributers to cardiac pathophysiology began to be better appreciated.

There are many types of cardiac morbidity. The major categories include dysrythmia, heart failure and ischemic heart disease. There are a number of ways that Major Depression contributes to those cardiovascular conditions. Some of the mechanisms by which Major Depression increases cardiovascular risk are the same as those in Metabolic Syndrome.

DYSRHYTHMIAS

Low heart rate variability (HRV) is a cardiovascular risk factor that is often seen in Major Depression[8]. Ordinarily, the heart changes the rate at which it beats quite often throughout the day. This reflects changing demands on the heart as well as moments of excitement, anger, or anxiety. Even breathing in and out subtly changes heart rate. When the heart rate doesn't change much throughout the day, there begins to be concern about whether the heart is somehow "stuck" in a certain rate. Low HRV is not a disorder itself, but rather it is a sign that something is amiss. Nonetheless, it is serious. Low HRV is associated with fatal dysrhythmias in depressed patients[9]. It is also likely to be accompanied by impairment of the baroreflex, which, in turn, is associated with increased risk of sudden cardiac death in patients with depression[10].

A common cause of low HRV is persistent activation of the sympathetic nervous system. The heart is essentially stuck in high gear. The sympathetic nervous system tends to be hyperactive in Major Depression[11]. This is known from studies showing increases in serum catecholamines in depressed patients. Sympathetic hyperactivity is further increased when depression is comorbid with an anxiety disorder. It is not clear if hyperactivity of the sympathetic nervous system causes depression, or if the agitation and dysphoria of depression drives the sympathetic nervous system. However, changes throughout the autonomic nervous system can occur in Major Depression, and the cause of low HRV in depressed patients is likely to be complex. For example, there is evidence that decreases in parasympathetic activity can contribute to low HRV in Major Depression[12]. In any case, HRV tends to normalize after successful treatment of depression[13].

As in depression, HRV is low in Metabolic Syndrome[14]. This is likely due to hyperactivity of the sympathetic nervous system. Whereas in Major

Depression this increase in sympathetic activity activation is psychogenic, in Metabolic Syndrome it is largely the result of hyperinsulinemia and inflammation. HRV normalizes in obese individuals after they begin to exercise and lose weight. This normalization of HRV is accompanied by reduction in signs of Metabolic Syndrome including improvements in insulin sensitivity, blood lipids, and glucose tolerance[15].

HEART FAILURE

Heart failure is simply the inability of the heart to maintain output of blood sufficient to meet the body's changing needs. It can be due to weakness of cardiac muscle following myocardial infarction or cardiomyopathy. Often, it is the manifestation of the heart's inability to overcome pressure in the vascular system, that is, hypertension. Hypertension is a major risk factor for heart disease, and an increase of only 10 points in the diastolic pressure increases the risk of heart disease by 37 percent[16].

Depression itself does not cause hypertension, but it can set the stage for its development. People with depression are half again as likely to develop hypertension over subsequent years than are people without depression[17]. Hypertension in depressed patients is likely due to the same overstimulation of the sympathetic nervous system that manifests as low HRV. A recent study has also shown that in the population of patients over 65 years old, Major Depression predisposes to heart failure[18]. Curiously, this increased risk of heart failure was seen in women but not men.

Although more attention has been paid to the role of Metabolic Syndrome in ischemic heart disease, the syndrome is known to be a risk factor for the development of heart failure[19]. In fact, in a study that followed men for 20 years, there was found to be an increase in the risk of left ventricular heart failure in the men that had signs of insulin resistance early in the study[20].

ISCHEMIC HEART DISEASE AND MYOCARDIAL INFARCTION

Ischemic heart disease arises when the coronary arteries that supply the heart with blood begin to narrow. The pathology primarily responsible for the narrowing of these arteries is atherosclerosis. Atherosclerosis is the accumulation of fatty material and plaques in the lining and inner layers of the coronary arteries. This narrowing usually develops slowly. Discomfort begins to be experienced during exertion when the needs of the myocardium outpace its blood supply. As the condition progresses, plaques in the coronary arteries grow more likely to ulcerate and become sites of clot formation. The end result can be myocardial infarction.

Atherosclerosis begins when injuries to the endothelial lining of the coronary arteries stimulate platelets and monocytes to adhere to the damaged tissue[21]. The damage need not be severe for this to occur. Hypertension, chemical insult from smoking, or even hyperlipidemia may be sufficient irritation to initiate this process. Cytokines released from the adhering platelets and mono-cytes stimulate migration of smooth muscle cells into the intima of the artery. Ostensibly, this is to help bolster the integrity of the arterial wall. The smooth muscle cells that gather within the arterial walls release proteins whose pur-pose is to strengthen and elasticize these structures. This collection of smooth muscle cells and protein matrix can serve as a nidus for accumulation of other substances, including cholesterol-rich serum.

The cytokines released by platelets, monocytes and damaged endothelium also trigger the maturation of monocytes into macrophages. The macrophages migrate into the intima to stand guard over the repair process. Macrophages are formidable soldiers in the immune system's army of cells. One of their tactics is to release strong oxidizing substances to destroy invading organ-isms. Unfortunately, this action by macrophages oxidizes LDL cholesterol that has seeped into the damaged tissue. Oxidized LDL is avidly taken up by macrophages. The macrophages gather up so much cholesterol they become what are referred to as foam cells. This accumulation of cholesterol in bloated foam cells is largely responsible for the build up of fatty material in damaged coronary arteries.

HDL helps prevent atherosclerosis. It does so by several mechanisms[22]. Foremost, it participates in the so-called reverse cholesterol transport system that carries free and LDL-bound cholesterol back to the liver for safe dis-posal. It also helps prevent the oxidation of LDL that makes it so appealing to hungry macrophages. Other beneficial effects of HDL include reducing thrombogenesis and improving the function of endothelium.

Recurrent episodes of Major Depression are associated with increased risk of atherosclerosis[23]. Anger, anxiety and hostility are often thought of as the psychological characteristics that contribute to hypertension and car-diovascular disease. However, hopelessness[24], discontent[25], anhedonia, and neurovegetative changes[26] such as fatigue and appetite disturbance, are more clearly linked with progression of atherosclerosis. Pathological changes known to exist in both Major Depression and Metabolic Syndrome are likely to contribute to atherosclerosis. Among these changes are platelet hyperactivity, inflammation, stress, and serum lipid abnormalities.

PLATELET HYPERACTIVITY

The reaction of platelets to endothelial injury, and the subsequent activation and adhesion of the platelets to the endothelium, are starting points in the pathogenesis of atherosclerosis. It stands to reason that if platelets are over-reactive, they are more likely to initiate this process. This, in turn, would enhance the likelihood of developing cardiovascular disease.

In people who are depressed, platelets are overly sensitive to the usual chemical triggers of activation[27]. There is enhancement of binding to P-selectin and glycoprotein IIb/IIIa in platelets of depressed patients. Elevations in serum levels of platelet-specific substances, such as platelet factor 4, show that platelets tend to remain in a state of activation in Major Depression.

Although there is evidence that platetes are hypersensitive, there is con-flicting evidence about whether platelets are more likely to aggregate and form clots in patients with depression[28]. Thrombin-induced platelet aggregation is reported to be enhanced in patients with depression. However, there is little evidence that serotonin-induced aggregation is altered. This is despite the fact that some effects of serotonin, such as the mobilization of intracellular calcium, are enhanced in the platelets of depressed patients.

It is possible that in patients with Major Depression, hyperactive platelets contribute to atherosclerosis and cardiovascular risk not by aggregation, but from other aspects of their activity. P-selectin is part of the mechanism by which platelets adhere to damaged endothelium. Because the activity of P-selectin is enhanced in Major Depression, there may be an increase in the likelihood that platelets adhere to endothelium and initiate the process leading to atherosclerosis. Platelet-to-platelet adhesion, that is, aggregation, does not appear to be enhanced in depressed patients. However, it remains to be determined if there are changes in platelet-to-endothelium adhesion in patients with Major Depression.

Platelet-induced inflammatory responses may also play a role in increasing cardiovascular risk in patients with depression. P-selectin initiates inflammatory processes in endothelium[29]. Other substances released by activated platelets, including platelet derived growth factor[30], soluble CD40 ligand[31], transforming growth factor-beta[32], and various cytokines, also stimulate inflammatory responses.

Unlike Major Depression, Metabolic Syndrome is clearly associated with increases in platelet aggregation. Some of the factors thought to enhance platelet aggregation in Metabolic Syndrome are increased levels of oxidative stress, hyperglycemia, obesity, high levels of leptin[33], low *omega-3* fatty acids[34], and insulin resistance[35]. It is possible that the adverse cardiovascular effects of Major Depression and Metabolic Syndrome are additive, and when they coexist, the risk of heart disease and death from myocardial infarction may increase.

INFLAMMATION

Inflammation contributes to irritation and injury of the endothelial lining of arteries. It is damage to the endothelium that initiates the adhesion of the platelets and monocytes. Inflammation increases oxidative stress and thus decreases the body's ability to prevent the oxidation of LDL. Inflammatory cytokines stimulate the cells that participate in the pathogenesis of atherosclerosis. This accelerates the laying down of plaques of cholesterol and scar tissue that predispose to blockage of coronary arteries. Major Depression is associated with increases in inflammatory cytokines, many of which are the same substances elevated in Metabolic Syndrome.

The source of inflammation in Major Depression is not entirely clear. Some of the inflammatory cytokines may come from hyperactive platelets. Substances released from platelets can serve as triggers for the release of other inflammatory substances. IL-6 and TNF-alpha are inflammatory cytokines that are often elevated in the serum of patients with Major Depression. Platelet derived factors might stimulate the release of IL-6 and TNF-alpha from leukocytes, which are the primary source of these cytokines. However, there are other sources of these substances, including adipocytes. Indeed, adipocytes are seen as a major source of these cytokines in Metabolic Syndrome. Because TNF-alpha can stimulate platelets, it may be partially responsible for the hyperactivity of platelets seen in depressed patients. In any case, IL-6[36] and TNF-alpha[37] are elevated in depression and are considered to be markers of cardiac risk. NFk-B, which recent studies have found to have enhanced activity in depressed patients, is a trigger of inflammatory activity that plays a major role in the pathogenesis of atherosclerosis. Activation of NFk-B may be a necessary step in its development.

Patients with severe depression also tend to have patterns of serum protein elevations consistent with the acute phase response[38]. The liver is the source of the acute phase reactants. It alters its protein production and increases its output of the acute phase proteins in response to stimulation by cytokines such as TNF-alpha and IL-1. Among the proteins it releases in the acute phase is CRP. CRP is elevated in Major Depression, and it has recently emerged as an independent risk factor for cardiovascular disease[39].

Although inflammation was not part of Reaven's original conception of Syndrome X, he included it as a component of his expanded notion of Insulin Resistance Syndrome. The Metabolic Syndrome is often accompanied by increases in serum levels of markers of inflammation. Inflammation can arise from the increases in visceral adiposity, oxidative stress, glycosylation damage from hyperglycemia, toxic effects of poorly controlled fatty acid metabolism, and hyperinsulinemia itself. Inflammation is thought to be at least partially responsible for the increased serum levels of plasminogen activator inhibitor-1 (PAI-1) seen in Metabolic Syndrome[40]. Plasminogen activator is the remarkable substance that emergency department physicians administer intravenously to reverse life-threatening coronary and cerebrovascular clots. PAI-1 does the opposite. Increases in PAI-1 are associated with atherosclerosis and myocardial infarction[41].

STRESS

A large percentage of patients with Major Depression exhibit hypercortisolemia. Some of the adverse effect of hypercortisolemia on cardiovascular risk is simply that it exacerbates Metabolic Syndrome. However, high levels of cortisol may have more immediate adverse effects. In depressed patients, cortisol levels are directly related to serum levels of endothelin-1[42]. Endothelin-1 participates in inflammatory effects in the endothelium that predispose to endothelial injury, vasospasm and myocardial infarction[43]. Cortisol also hinders the synthesis of nitric oxide in coronary artery endothelium[44]. This would further place the heart at risk for vasospasm and infarction. Cortisol has long been associated with increased risk for atherosclerosis[45]; however, the mechanism remains controversial[46]. Association between cortisol and atherosclerosis may be best explained by cortisol's participation in Metabolic Syndrome.

LIPID ABNORMALITIES

High serum triglycerides and LDL accompanied by low HDL are fundamental components of the Metabolic Syndrome. They are a direct result of insulin resistance and were seen by Jerry Reaven as being a major source of cardiovascular risk in his original conception of Syndrome X. The higher the levels of LDL cholesterol, the more likely that macrophages will oxidize and engulf it. This process transforms macrophages into the swollen, cholesterol-laden foam cells that contribute to the development of atherosclerosis. Although triglycerides are known to cause intracellular damage in muscle, pancreas and liver in Metabolic Syndrome, it is not yet known what role they play in stimulating atherosclerosis. However, triglyceride-rich lipoproteins are suspected of playing a role in facilitating its development[47]. The low HDL in Metabolic Syndrome robs the body of some of its best mechanisms to slow the progression of atherosclerosis and coronary artery disease.

In Major Depression, as in Metabolic Syndrome, HDL cholesterol is often low[48]. In fact, HDL has in some studies been reported to be lowest in patients with severe, melancholic depression and in those that have histories of attempted suicides. It is likely that low HDL contributes to cardiovascular morbidity in depressed patients as it does in Metabolic Syndrome. Patients with the most severe Major Depression have also been found to have reductions in the activity of an important antioxidant enzyme called paraoxanase (PON 1)[49]. PON 1 helps prevent oxidation of both HDL and LDL, and it has been thought that reduced levels of PON 1 may be an independent risk factor for cardiovascular disease[50]. PON 1 activity is also reduced in Metabolic Syndrome[51]. Whether the reduction in PON 1 activity in patients with severe depression is related to Metabolic Syndrome or independent of it is not known.

Because of the tendency of patients with Major Depression to develop Metabolic Syndrome, one might suspect that these patients would also have elevated triglyceride and LDL levels. However, this is often not the case. That triglyceride and LDL levels are not as high as might be expected in Major Depression may be the result of the hypercortisolemia that often accompanies the illness[52]. Unfortunately, since hypercortisolemia itself may contribute to endothelial dysfunction and other aspects of cardiovascular disease, the relative decrease in triglycerides and LDL that it awards depressed patients gains little for them.

It is worth noting that cortisol itself may not be what is responsible for the blunting of hyperlipidemia in depressed patients. In studies of healthy young men, the administration of adrenocorticotrophic hormone (ACTH), but not cortisol, decreases serum levels of triglycerides and LDL[53]. Thus it is possible that the ACTH released by the pituitary gland to stimulate production and release of cortisol from the adrenal glands is what minimizes hyperlipidemia in some patients with Major Depression. What isn't clear is what occurs as the hypercortisolemia in depressed patients begins to stimulate the deposition of fat in visceral adipocytes. Visceral adipocytes possess the enzyme 11-HSD that regenerates cortisol out of less active cortisone and thus increases serum levels of cortisol. Ordinarily, high levels of cortisol feed back and reduce release of ACTH by the pituitary. However, in Major Depression, the negative feedback system in the hypothalamic-pituitary-adrenal axis may be faulty[54]. The question of whether hyperlipidemia becomes more or less likely in hyper-cortisolemic depressed patients with visceral obesity remains to be answered. Certainly, patients with Major Depression can have hyperlipidemia. In fact, the presence of hyperlipidemia in depressed patients has been associated with treatment resistance[55].

DOES TREATING MAJOR DEPRESSION REDUCE CARDIAC RISK?

If depression increases the risk of dying from heart disease, then it is reasonable to expect that treating depression in patients with heart disease would reduce that risk. Two major studies, ENRICHD[56] and SADHART[57] have been performed to evaluate the benfits of treating depressed post-myocardial infarction (MI) patients with psychotherapy and/or antidepressants. The results have been less than spectacular. In the ENRICHD study post-MI patients received cognitive behavioral therapy with or without setraline. The only benefits of the interventions were what were described as modest in improvements in quality of life. There was no decrease in the likelihood of a second MI. In SADHART, there were mere suggestions that sertraline might help prevent a second event in post-MI patients. Among patients treated with sertraline, 14 percent had a second MI, whereas among those that received the placebo, 22 percent had a second MI. Even more disappointing in SADHART was the fact that ser-traline was not significantly more effective than the placebo in treating the depression of these patients. A smaller study performed at roughly the same time as ENRICHD and SADHART did show that adding an antidepressant, primarily an SSRI, could reduce the risk of a second MI by nearly one-half[58]. It is not clear why the results of these studies differed.

Unfortunately, the relationship between treating depression with antide-pressants and reducing cardiac risk is difficult to study. This is the case because many of the antidepressants, particularly the SSRIs, block some of the effects of platelet activation that likely contribute to cardiac risk[59]. Thus, it is possi-ble that SSRIs reduce cardiac risk through a mechanism that has nothing to do with depression itself. Indeed, in the ENRICHD study, in cases in which Cognitive Behavioral Therapy was used without drugs to relieve depression in patients after their heart attack, the risk of dying from a second heart attack was not reduced[60].

Patients with a strong history of Major Depression are at increased risk for heart disease even if they become euthymic. This could mean that the adverse effects of depression on the heart are due to depression being a trait rather than a state. On the other hand, it might simply mean that once a person has suffered severe depression, damage to the cardiovascular system has already occurred. In either case, treating the biological manifestations of Major Depression is the most prudent course to take when trying to reduce cardiac risk. This would be aimed at treating the cardiovascular risk factors that Major Depression shares with Metabolic Syndrome, such as hypertension, sympathetic nervous system activation, and low HDL. It should include addressing the hyperlipidemia that characterizes Metabolic Syndrome and the hypercortisolemia that so often occurs in patients with depression. Medications can be tailored to address these problems. However, the conscientious physician will pay attention to healthy diet, adequate sleep, sufficent exercise, and reduction of stress to help correct these problems.

REFERENCES

1. Weeke, A., Causes of death in manic-depressives. In *Origin Prevention and Treatment of Affective Disorders*, M. Schou and E. Stromgren, eds. London, Academic Press, 1979, pp. 289–299.
2. Carney, R.M. et al., Major depressive disorder predicts cardiac events in patients with coronary artery disease. *Psychosom. Med.* 1988; 50:627–633.
3. Giltay, E.J. et al., Dispositional optimism and all-cause and cardiovascular mortality in a prospective cohort of elderly Dutch men and women. *Arch. Gen. Psychiatry* 2004; 61:1126–1135.
4. Glassman, A.H. and Shapiro, P.A., Depression and the course of coronary artery disease. *Am. J. Psychiatry* 1998; 155:4–11.
5. Rudisch, B. and Nemeroff, C.B., Epidemiology of comorbid coronary artery disease and depression. *Biol. Psychiatry* 2003; 54:227–240.
6. Hasin, D.S. et al., Epidemiology of major depressive disorder: results from the National Epidemiologic Survey on Alcoholism and Related Conditions. *Arch. Gen. Psychiatry* 2005; 62:1097–1106.
7. Barth, J., Schumacher, M., and Herrmann-Lingen, C., Depression as a risk factor for mortality in patients with coronary heart disease: a meta-analysis. *Psychosom. Med.* 2004; 66:802–813.
8. Musselman, D.L., Evans, D.L., and Nemeroff, C.B., The relationship of depression to cardiovascular disease: epidemiology, biology, and treatment. *Arch. Gen. Psychiatry* 1998; 55:580–592.
9. Agelink, M.W. et al., A functional-structural model to understand cardiac autonomic nervous system (ANS) dysregulation in affective illness and to elucidate the ANS effects of antidepressive treatment. *Eur. J. Med. Res.* 2004; 9:37–50.
10. Davydov, D.M. et al., Baroreflex mechanisms in major depression. *Prog. Neuropsychopharmacol. Biol. Psychiatry* 2007; 31:164–177.
11. Veith, R.C. et al., Sympathetic nervous system activity in major depression. Basal and desipramine-induced alterations in plasma norepinephrine kinetics. *Arch. Gen. Psychiatry* 1994; 51:411–422.
12. Yeragani, V.K. et al., Diminished chaos of heart rate time series in patients with major depression. *Biol. Psychiatry* 2002; 51:733–744.
13. Balogh, S. et al., Increases in heart rate variability with successful treatment in patients with major depressive disorder. *Psychopharmacol. Bull.* 1993; 29(2):201–206.
14. Liao, D. et al., Cardiac autonomic function and incident coronary heart disease: a population-based case-cohort study. The ARIC Study. Atherosclerosis Risk in Communities Study. *Am. J. Epidemiol.* 1997; 145:696–706.
15. Ito, H. et al., Effects of increased physical activity and mild calorie restriction on heart rate variability in obese women. *Jpn Heart J.* 2001; 42:459–469.
16. Wilson, P.W. et al., Prediction of coronary heart disease using risk factor catagories. *Circulation* 1998; 97:1837–1847.
17. Jonas, B.S. and Lando, J.F., Negative affect as a prospective risk factor for hypertension. *Psychosom. Med.* 2000; 62:188–196.
18. Williams, S.A. et al., Depression and risk of heart failure among the elderly: a prospective community-based study. *Psychosom. Med.* 2002; 64:6–12.
19. Ingelsson, E. et al., Metabolic syndrome and risk for heart failure in middle-aged men. *Heart* 2006; 92:1409–1413.
20. Arnlov, J. et al., Several factors associated with the insulin resistance syndrome are predictors of left ventricular systolic dysfunction in a male population after 20 years of follow-up. *Am. Heart J.* 2001; 142:720–724.

21. Steinberg, D., Current theories of the pathogenesis of atherosclerosis. In *Hypercholesterolemia and Atherosclerosis, Pathogenesis and Prevention*, D. Steinberg and J.M. Olefsky, eds. New York, Churchill-Livingstone, 1987, p. 5.

22. Choi, B.G. et al., The role of high-density lipoprotein cholesterol in atherothrombosis. *Mt. Sinai J. Med.* 2006; 73:690–701.

23. Jones, D.J. et al., Lifetime history of depression and carotid atherosclerosis in middle-aged women. *Arch. Gen. Psychiatry* 2003; 60:153–160.

24. Everson, S.A. et al., Hopelessness and 4-year progression of carotid atherosclerosis. The Kuopio Ischemic Heart Disease Risk Factor Study. *Arterioscler. Thromb. Vasc. Biol.* 1997; 17:1490–1495.

25. Agewall, S. et al., Negative feelings (discontent) predict progress of intima-media thickness of the common carotid artery in treated hypertensive men at high cardiovascular risk. *Am. J. Hypertens.* 1996; 9:545–550.

26. Stewart, J.C. et al., Negative emotions and 3-year progression of subclinical atherosclerosis. *Arch. Gen. Psychiatry* 2007; 64:225–233.

27. Musselman, D.L. et al., Exaggerated platelet reactivity in major depression. *Am. J. Psychiatry* 1996; 153:1313–1317.

28. Mendelson, S.D., The current status of the platelet 5-HT(2A) receptor in depression. *J. Affect. Disord.* 2000; 57:13–24.

29. Egami, K. et al., Ischemia-induced angiogenesis: role of inflammatory response mediated by P-selectin. *J. Leukoc. Biol.* 2006; 79:971–976.

30. Cheon, H. et al., Platelet-derived growth factor-AA increases IL-1beta and IL-8 expression and activates NF-kappaB in rheumatoid fibroblast-like synoviocytes. *Scand. J. Immunol.* 2004; 60:455–462.

31. Blumberg, N. et al., An association of soluble CD40 ligand (CD154) with adverse reactions to platelet transfusions. *Transfusion* 2006; 46:1813–1821.

32. Singh, N.N. and Ramji, D.P., The role of transforming growth factor-beta in atherosclerosis. *Cytokine Growth Factor Rev.* 2006; 17:487–499.

33. Nakata, M. et al., Leptin promotes aggregation of human platelets via the long form of its receptor. *Diabetes* 1999; 48:426–429.

34. Dunn, E.J. and Grant, P.J., Type 2 diabetes: an atherothrombotic syndrome. *Curr. Mol. Med.* 2005; 5:323–332.

35. Ferroni, P. et al., Platelet activation in type 2 diabetes mellitus. *J. Thromb. Haemost.* 2004; 2:1282–1291.

36. Koenig, W. et al., Increased concentrations of C-reactive protein and IL-6 but not IL-18 are independently associated with incident coronary events in middle-aged men and women: results from the MONICA/KORA Augsburg case-cohort study, 1984–2002. *Arterioscler. Thromb. Vasc. Biol.* 2006; 26:2745–2751.

37. Tuomisto, K. et al., C-reactive protein, interleukin-6 and tumor necrosis factor alpha as predictors of incident coronary and cardiovascular events and total mortality. A population-based, prospective study. *Thromb. Haemost.* 2006; 95:511–518.

38. Maes, M. et al., Disturbances in acute phase plasma proteins during melancholia: additional evidence for the presence of an inflammatory process during that illness. *Prog. Neuropsychopharmacol. Biol. Psychiatry* 1992; 16:501–515.

39. Arena, R. et al., The relationship between C-reactive protein and other cardiovascular risk factors in men and women. *J. Cardiopulm. Rehabil.* 2006; 26:323–327.

40. Alessi, M.C. and Juhan-Vague, I., Contribution of PAI-1 in cardiovascular pathology. *Arch. Mal. Coeur. Vaiss.* 2004; 97:673–678.

41. Bonora, E., The metabolic syndrome and cardiovascular disease. *Ann. Med.* 2006; 38:64–80.

42. Lederbogen, F. et al., Endothelin-1 plasma concentrations in depressed patients and healthy controls. *Neuropsychobiology* 1999; 40:121–123.

43. Ihling, C. et al., Endothelin-1 and endothelin converting enzyme-1 in human atherosclerosis – novel targets for pharmacotherapy in atherosclerosis. *Curr. Vasc. Pharmacol.* 2004; 2:249–258.

44. Rogers, K.M. et al., Inhibitory effect of glucocorticoid on coronary artery endothelial function. *Am. J. Physiol. Heart Circ. Physiol.* 2002; 283:H1922–1928.

45. Troxler, R.G. et al., The association of elevated plasma cortisol and early atherosclerosis as demonstrated by coronary angiography. *Atherosclerosis* 1977; 26:151–162.

46. Tedeschi-Reiner, E. et al., Plasma cortisol in men–relationship with atherosclerosis of retinal arteries. *Coll. Antropol.* 2002; 26:615–619.

47. Tanaka, A. et al., Metabolism of triglyceride-rich lipoproteins and their role in atherosclerosis. *Ann. N.Y. Acad. Sci.* 2001; 947:207–212.

48. Maes, M. et al., Lower serum high-density lipoprotein cholesterol (HDL-C) in major depression and in depressed men with serious suicidal attempts: relationship with immune-inflammatory markers. *Acta Psychiatr. Scand.* 1997; 95:212–221.

49. Sarandol, A. et al., Oxidation of apolipoprotein B-containing lipoproteins and serum paraoxanse/arylesterase activities in major depressive disorder. *Prog. Neuropsychopharmacol. Biol. Psychiatry* 2006; 30:1103–1108.

50. Mackness, B. et al., Low paraoxanase activity predicts coronary events in the Caerphilly Prospective Study. *Circulation* 2003; 107:2775–2779.

51. Sentí, M. et al., Antioxidant paraoxonase 1 activity in the metabolic syndrome. *J. Clin. Endocrinol. Metab.* 2003; 88:5422–5426.

52. Kopf, D. et al., Lipid metabolism and insulin resistance in depressed patients: significance of weight, hypercortisolism, and antidepressant treatment. *J. Clin. Psychopharmacol.* 2004; 24:527–531.

53. Berg, A.L. et al., The effects of adrenocorticotrophic hormone and an equivalent dose of cortisol on the serum concentrations of lipids, lipoproteins, and apolipoproteins. *Metabolism* 2006; 55:1083–1087.

54. Plotsky, P.M., Owens, M.J., and Nemeroff, C.B., Psychoneuroendocrinology of depression. Hypothalamic-pituitary-adrenal axis. *Psychiatr. Clin. North Am.* 1998; 21:293–307.

55. Papakostas, G.I. et al., Serum cholesterol in treatment-resistant depression. *Neuropsychobiology* 2003; 47:146–151.

56. Berkman, L.F. et al., Effects of treating depression and low perceived social support on clinical events after myocardial infarction: the Enhancing Recovery in Coronary Heart Disease Patients (ENRICHD) Randomized Trial. *JAMA* 2003; 289:3106–3116.

57. Glassman, A.H. et al., Sertraline antidepressant heart attack randomized trial (SADHEART) group. Sertraline treatment of major depression in patients with acute MI or unstable angina. *JAMA* 2002; 288:701–709.

58. Sauer, W.H., Berlin, J.A., and Kimmel, S.E., Effect of antidepressants and their relative affinity for the serotonin transporter on the risk of myocardial infarction. *Circulation* 2003; 108:32–36.

59. Serebruany, V.L. et al., Relationship between release of platelet/endothelial biomarkers and plasma levels of sertraline and N-desmethylsertraline in acute coronary syndrome patients receiving SSRI treatment for depression. *Am. J. Psychiatry* 2005; 162:1165–1170.

60. Writing Committee for the ENRICHD Investigators, Effects of treating depression and low perceived social support on clinical events after myocardial infarction. The enhancing recovery in coronary heart disease patients (ENRICHD) randomized trial. *JAMA* 2003; 289:3106–3116.

7

METABOLIC SYNDROME, INSULIN AND ALZHEIMER'S DISEASE

In 1907, a distraught husband brought his 51-year-old wife to see the German Neurologist, Alois Alzheimer. Over a period of only a few years, she had developed problems with her memory and ability to perform her usual daily functions. By the time he brought her to Dr. Alzheimer, this once charming woman was exhibiting behaviors completely uncharacteristic of her usual way of acting. She was insanely jealous of her husband. She had fits of screaming and hallucinations. Dr. Alzheimer had nothing to offer the unfortunate woman. She was placed in an asylum, and not too long afterward she died there. However, Dr. Alzheimer had been so struck by the rapid and precipitous decline in the woman's function, that he had continued to observe her. When she died he removed her brain for pathological study.

The year 1907 may seem a time too distant in the past to provide us with any useful scientific information. However, in 1906, the Nobel prize for Medicine had been awarded to two of the greatest neuroscientists of the nineteenth and, perhaps, twentieth centuries, Camillo Golgi and Santiago Ramon y Cajal. Golgi had developed a method to stain brain tissue that is still in use today. His process almost magically turned only a small percentage of individual neurons jet black under the microscope. It is still not entirely clear why his process does not simply turn the entire slice of brain and all the neurons it contained black. Thankfully it does not, and using Golgi's stain, Ramon y Cajal sat for years in front of his microscope, diligently, artfully and with amazing precision, drawing pictures and diagrams of what normal neurons from all the different areas of the human brain look like.

Because of the work of Golgi and Ramon y Cajal, Alzheimer had very good information about what normal tissue looks like under the microscopic. The brain of Alzheimer's patient was remarkably shrunken. However, the microscopic examination of her brain tissue was the most revealing part of his evaluation. The tissue was grossly abnormal. Alzheimer described catastrophic loss of normal neurons, and replacement with unusual plaques and tangles.

Because of the relatively young age of Alzheimer's patient, and the rapid, precipitous decline she exhibited, it is likely that she suffered an extreme form of Alzheimer's Disease now referred to as early onset Alzheimer's dementia. That unfortunate woman's presentation is not typical of the disease. More commonly, the illness begins in a patient's late 60s, and slowly develops over many years' time. Generally, it is only in the last few years of the illness that it progresses rapidly. In fact, most cases of Alzheimer's are not even diagnosed until they have progressed to the moderately severe stage. This may be as long as 10 years after the disease is likely to have actually started.

THE STAGES OF ALZHEIMER'S DISEASE

The first symptoms of Alzheimer's generally become apparent in dealings with the outside world. People have difficulty learning new things. If they are still working, their performance is seen to be deteriorating. They make mistakes and get confused doing things they may have done for years. They get behind in paying their bills. Checks bounce. They are forgetful. Names of acquaintances are easily forgotten. They lose things and forget where the car is parked. The day comes when they get lost on the way to the grocery store where they have driven hundreds of times before. It is quite common for anxiety and depression to evolve at this time, when the sufferer is aware enough to know that something is very wrong with them.

As the disease progresses to the moderate stage, the deficits begin to appear within the home sphere. Food rots in the refrigerator. Dishes pile up in the sink. Hygiene suffers. The stove can be left on, or food gets burned when food preparation is abandoned in midcourse. By this time, memory is very obviously deteriorating. Sufferers do not recall events that may have occurred earlier that day. Statements are repeated over and over. It is difficult to find words for common objects, and names of even close friends or relatives are forgotten. In fortunate cases, the patient experiences a slide into serene befuddlement. However, severe anxiety, outbursts of anger, or outright aggression can occur. It is not uncommon for sufferers to begin to imagine things. The favorite bowl they mistakenly threw in the trash is thought to have been stolen by neighbors. Concerned family members may be seen as trying to control them or take their house or savings away. At this stage, families begin to wonder about their loved one's ability to function in their home. They may begin to consider assisted living or nursing home placement.

In the final stages of Alzheimer's Disease, control of the body itself is lost. Speech deteriorates and patients have great difficulty expressing themselves. Some call out for help, even when help is right beside them. Others scream for no obvious reason, or they become completely mute. Sufferers generally become incontinent of bladder and bowels. They are unable to feed themselves. Delusions and emotional outbursts are joined by delirium and hallucinations. Patients may be oriented to self, but they lose conception of time and place.

Even spouses and children may no longer be recognized. Walking becomes more difficult, and a point is reached when patients become bed-bound or must be restrained to prevent falls. Alzheimer's itself does not kill, but death often comes from combinations of weakness, dehydration, failure to eat, and pneumonia. The pneumonia is often the result of aspiration of food or fluid, and from inability to guard the airway and generate an adequate cough reflex.

PLAQUES AND TANGLES

Alzheimer described the plaques and tangles in the brains of sufferers of the disease that was eventually named after him. However, he lacked the techniques needed to identify the chemical nature of these microscopic lesions. Modern research has determined that the plaques are formed from what is essentially crystallization of an abnormal protein called amyloid. The tangles are knots of the protein *tau*, which is a substance normally found in neurons. *Tau* helps form the scaffolding that gives neurons their shape, and it helps form a transport system for carrying other substances throughout the nerve cell.

Although many factors contribute to the development and progression of Alzheimer's Disease, the laying down of amyloid plaque in the brain is the primary problem in the illness. No one knows exactly why this occurs. Some studies have shown that the accumulation of amyloid protein is due to increase in the production of the precursor to amyloid, which is simply known as amyloid precursor protein (APP). Others have suggested that improper cleavage of APP produces an abnormal, sticky form of amyloid that the body cannot process properly. Some researchers point to deficits in the body's ability to disperse and carry away the amyloid before it accumulates and "crystallizes". In fact, there is evidence that all three pathological processes may contribute to Alzheimer's Disease.

INSULIN AND ALZHEIMER'S DISEASE

Of relevance to the present discussion is the growing evidence that Metabolic Syndrome exacerbates and accelerates the progression of Azheimer's. The risk of developing Alzheimer's Disease is nearly doubled in people with adult onset diabetes in comparison with those without this condition[1]. Generally, adult onset diabetes, which is Diabetes Type II, arrives at the end of years of suffering Metabolic Syndrome. A number of researchers are coming to suspect that the insulin resistance and subsequent hyperinsulinemia that occurs in Metabolic Syndrome is at least partially responsible for this increase in Alzheimer's[2]. In fact, hyperinsulinemia can double the risk of Alzheimer's Disease[3].

In a recent study[4], mice genetically altered to produce human brain proteins were made insulin resistant and hyperinsulinemic by dietary changes. In fundamental respects, their condition resembled Metabolic Syndrome in humans. The brains of these animals were found to have increases in activity of gamma-secretase, the enzyme that cleaves APP into amyloid, as well as increases in amyloid plaque burden. Perhaps most importantly, the learning ability of these animals was impaired. In humans, hyperinsulinemia increases the amount of the sticky, 1-42 form of amyloid in the blood and cerebrospinal fluid. This is the form of amyloid that builds up in the brains of Alzheimer's patients[5].

In what ways might an impairment of insulin's activity affect the brain? Although the brain's primary fuel is glucose, brain tissue does not depend on the action of insulin to enhance its uptake of glucose. The transport molecules that carry glucose across the cell membranes of neurons in most areas of the brain are Type I transporters. This type of glucose transporter does not require insulin to be active in taking up glucose. For many years it was assumed that the brain was indifferent to ambient insulin levels. Therefore, the scientific world was surprised in 1967 when insulin was found in fairly high concentration in the cerebrospinal fluid that bathes the brain[6]. In 1978, insulin receptors were identified in the brain[7]. There have subsequently been descriptions of special transport molecules that shuttle insulin from the blood across the blood brain barrier into brain tissue[8]. There has even been a controversial claim that the brain itself makes insulin for its own special purposes[9]. In view of knowledge that insulin was not needed for glucose metabolism, the question arose as to what functional role insulin might serve in the brain.

It turns out that insulin is important for a variety of brain activities. Not surprisingly, one function of insulin is control of food intake, weight regulation, and energy metabolism. Insulin receptors in the hypothalamus mediate these effects. However, the hypothalamus is an old and primitive part of the brain that controls basic physiological functions. It was more difficult to explain the finding of insulin and insulin receptors in areas of the brain involved in higher brain functions. For example, in rat brain, insulin receptors have also been found in the hippocampus, pyriform cortex, olfactory bulbs and cerebellum[10]. These areas of the brain have little to do with food intake or metabolism.

Studies in which insulin receptors have been deleted from mice through the gene "knock out" method have shown no discernible deficits in learning or memory in those animals[11]. However, injection of the antibiotic strepto-zotocin directly into rat brain caused severe neurodegeneration quite similar to that seen in Alzheimer's Disease[12]. Streptozotocin destroys insulin produc-ing cells in the pancreas, and peripheral administration is commonly used to induce diabetes in experimental animals. Administration of PPAR-delta ago-nists decreased amyloid production and reversed many of the learning defects observed in these streptozotocin-injected animals. It was suspected that the capacity of the brain to produce its own insulin was eliminated by the antibi-otic, and the subsequent neurodegenerative damage was due to depletion of intraneuronal insulin.

There are studies suggesting that insulin can act in the human brain to enhance cognitive function. The olfactory bulbs are areas of the brain rich in insulin receptors. Some of the nerve fibers sent out from the olfactory bulbs end up in the hippocampus[13]. The hippocampus plays an important role in learning and memory. Intranasal administration of insulin has been found to enhance memory in humans. This appears to be the result of transport of insulin through the nasal mucosa, into the olfactory nerves, and on to the olfactory bulbs. This improvement of memory is not due to effects on serum levels of insulin or glucose, as these did not change with intranasal administration[14]. In fact, increasing serum insulin while maintaining serum glucose at normal levels, through a technique known as glucose clamping, also enhances human memory and cognitive function[15]. Thus insulin itself appears to improve cognitive function.

INSULIN-DEGRADING ENZYME AND AMYLOID

There is evidence that in people with insulin resistance, a state of resistance to insulin can exist in the brain as well as in the peripheral tissues. Thus, if lack of insulin causes deficits in cognitive function, then the resistance to insulin seen in Metabolic Syndrome might be expected to diminish cognition by blunting the effects of insulin in the brain. The compensatory increases in insulin levels that occur in Metabolic Syndrome may also cause damage to the brain. Interestingly, some of the damage from hyperinsulemia may not be due to insulin itself, but rather to the way the body destroys insulin after it has served its purpose.

It is generally the case that chemical messengers of the body, such as insulin, are destroyed after they have delivered their message. If this were not the case, the message would be prolonged and imprecise. If insulin were not readily destroyed, levels of the hormone would remain high in the blood and continue to lower glucose levels even after those levels had been brought back to normal. In 1959, Dr. Henry Tomizawa, a biochemist at the University of Washington School of Medicine, reported his finding of an enzyme in beef liver that destroyed insulin[16]. This enzyme has since been referred to as insulin degrading enzyme (IDE). It is now clear that IDE is one of the primary ways in which the body controls the effects of insulin.

As is often the case, when an enzyme does its work well, the body finds additional uses for it. It is here that another connection between insulin and Alzheimer's Disease becomes apparent. IDE not only breaks down insulin, but it is also one of the more important enzymes in the breakdown of amyloid[17]. There is now compelling evidence that as levels of insulin rise in the brain, insulin competes with amyloid for access to IDE. In simple terms, the more insulin IDE has to breakdown, the less amyloid it is able to destroy. Thus in Metabolic Syndrome, where insulin levels are high in the body's attempt to overcome insulin resistance, the brain must struggle harder to rid itself of amyloid.

The insulin resistance that occurs in Metabolic Syndrome may make hyper-insulinemia even more of a threat to the brain. When insulin stimulates the insulin receptor, one of the downstream effects is an increase in production of IDE. This is a feed-forward effect quite commonly seen in physiology. Under normal circumstances, stimulation of IDE synthesis is the way that insulin puts the brakes on its own activity. In the state of insulin resistance, insulin does not adequately stimulate its receptor, and one consequence may be that the IDE that would have been synthesized to destroy the insulin is no longer produced in adequate quantity. Thus, the quantity of insulin might remain abnormally high. There would be less IDE to break apart amyloid, and more insulin competing with amyloid for the little IDE that is actually there.

The high levels of stress hormone in the blood often found in Metabolic Syndrome further reduce the availability of IDE. Studies evaluating the effects of cortisol on the brains of macaque monkeys showed that year-long exposure to high levels of the stress hormone caused decreases in production of IDE. Stress also increases the ratio of sticky amyloid 1-42 to the normal, less sticky 1-40 form of the protein[18].

As if those problems were not enough, there is yet one more factor in the upward spiral of pathological changes that arises from the structural similarities of amyloid and insulin. Not only are insulin and amyloid similar enough in structure to compete for sites on the IDE molecule that destroys them, both also compete for binding to the insulin receptor[19]. The insulin resistance that increases amyloid production results in even more, albeit, a different type of insulin resistance due to insulin then having to compete with that amyloid for access to the insulin receptor.

INSULIN AND TANGLED TAU

The pathological changes in *tau* protein, which forms the tangles first described by Alzheimer, are generally thought to occur after the accumulation of amyloid plaque has disrupted neuronal activity. However, insulin may directly affect *tau* and help prevent a process called hyperphosphorylation that disrupts tau's structure and activity. *Tau* is far more readily hyperphosphorylated in knockout mice genetically engineered to have defective brain insulin receptors[20]. Insulin is also known to inhibit activity of the enzyme glycogen synthase kinase 3 (GSK-3). GSK-3 phosphorylates *tau*, and inhibition of its activity by insulin prevents hyperphosphorylation[21]. Thus, it is reasonable to suspect that insulin resistance in the brain, and the resulting decrease in the effects of insulin, might lead to increases in neurofibrillary tangles of *tau* protein. Amyloid protein, which competes with insulin in several important ways, happens to stimulate GSK-3[22]. Thus, as abnormal amyloid protein builds up in brain tissue, the hyperphosphorylation of *tau* is further enhanced.

Some authors suggest that resistance of brain tissue to the effects of insulin contributes to the development of Alzheimer's Disease. However, there are reports that some Alzheimer's patients have low CSF-to-plasma ratios of insulin[23]. In chronic hyperinsulinemia, such as occurs in Metabolic Syndrome, insulin levels in CSF can decrease due to down-regulation of insulin transport across the blood-brain barrier. Thus, it is possible that Metabolic Syndrome can in some cases contribute to Alzheimer's Disease by depriving the brain of sufficient insulin.

As noted previously, the lithium used in treatment of Bipolar Affective Disorder inhibits GSK-3. In fact, lithium has been found to be useful in treating dementia secondary to HIV infection. In HIV-induced dementia, *tau* protein is hyperphosphorylated and tangleed, as it is in Alzheimer's Disease. In view of the fact that many medications useful in the treatment of BPAD are also inhibitors of GSK-3 activity, it is possible that some of the agitation and behavioral dyscontrol seen in Alzheimer's Disease is due to disinhibition of GSK-3. Medications that inhibit GSK-3, such as valproate[24], lithium[25], can be useful in controlling agitation in Alzheimer's patients. Atypical neuroleptics, which inhibit GSK-3 activity, can also help relieve the agitation of Alzheimer's Disease[26]. However, side effects such as postural hypotension and somnolence can limit their usefulness.

INFLAMMATION AND ALZHEIMER'S DISEASE

Inflammation is another process in the brain that contributes to the initiation and progression of Alzheimer's Disease. At one time the brain was thought to be free of immune system cells. However, it is now known that some of the glial cells that exist around the neurons of the brain serve several purposes. The glial cells usually perform housekeeping functions in the brain. Some help mop up used neurotransmitters and clean up after neurons that die. Others wrap themselves around nerve fibers and form insulation to accelerate and focus nerve conduction. However, when inflammation starts inside the brain, such as from oxidative damage or irritation from collections of amyloid protein, the glial cells can begin to function like macrophages. They can release a number of powerful substances that can do good, or harm if out of control. When amyloid takes up permanent residence in the brain tissue, so do the transformed glial cells.

The role of inflammation in Alzheimer's became a major focus of attention after reports that people with long histories of treatment with non-steroidal anti-inflammatory drugs (NSAIDs) were less likely to develop Alzheimer's[27]. The NSAIDs include many common drugs such as ibuprofen. This class of drugs acts primarily by blocking the production of inflammatory prostaglandins from arachidonic acid. Unfortunately, controlled studies of the effects of NSAIDs on Alzheimer's have been somewhat disappointing. Not all NSAIDs seem to provide protection from the development of Alzheimer's. Indeed, it is now suspected, that benefits of particular NSAIDs, such as ibuprofen and indomethacin, may be due to effects other than inhibition of prostaglandin protection. In fact, ibuprofen is also known to reduce production of amyloid, enhance activity of PPAR, and reduce the activity of beta-secretase, an enzyme that splices APP into amyloid[28]. Unfortunately, ibuprofen does not readily pass through the blood brain barrier, thus the doses needed to produce the desired effects in the brain might not be tolerated by the rest of the body. Because of this fact, there are currently human trials of other "profens" similar to ibuprofen in structure and activity but better able to enter brain tissue.

Although the jury is still out on the question of prostaglandin involvement in Alzheimer's, there is still considerable evidence that other inflammatory pathways are contributing factors in the disease. A recent review by Ringheim and Szczepanik[29] provides an excellent overview of current understanding of inflammation and Alzheimer's. They described a process in which small bits of amyloid protein in the brain generate an immune response mediated by glial cells. The glial cells release cytokines that further stimulate production of amyloid. That amyloid, in turn, causes further release of cytokines. This positive feedback system eventually spirals out of control.

Earlier, I had mentioned the fact that the brain is insulated from effects of substances in the blood by the so-called blood-brain barrier (BBB). Thus, it is reasonable to wonder if the brain is affected by the chemical mediators of inflammation that are increased in the blood in Metabolic Syndrome. The answer is yes, and no. Some inflammation-inducing substances damage the BBB and make it leak, which provides otherwise barred substances access to the brain. However, some inflammatory substances in the blood can bind to the lining of capillaries and, without themselves passing through it, trigger inflammatory changes on the other side. Some inflammatory cytokines, including TNF and IL-6, bind to capillary walls that are technically outside the brain, and cause inflammatory prostaglandins to be released into brain tissue[30]. Similarly, if broken-up bits of bacteria are injected into rats, even in doses too small to damage the BBB, there are still increases in the inflammatory cytokines TNF-alpha, IL-1 and IL-6 in their brain tissue[31]. A study using transgenic mice found that injection of bacterial wall fragments into the blood causes both immune activation and increases in amyloid deposition in brain tissue[32]. This is the same deposition of amyloid that causes Alzheimer's Disease.

Another strong indication of relationships between Metabolic Syndrome, inflammation and Alzheimer's Disease in humans comes from a 2004 study by Dr. Kristine Yaffe[33] from the University of California at San Francisco. In her study, men in their 70s were evaluated for having Metabolic Syndrome, and whether or not they showed indications of having ongoing inflammatory processes in their bodies. The presence of Metabolic Syndrome was defined as having abdominal obesity; elevated serum triglyceride levels, low HDL, hypertension; and elevated fasting glucose. The presence of actual diabetes disqualified potential participants, thus the abnormalities in physiology and energy metabolism were due solely to Metabolic Syndrome. The existence of inflammatory processes was defined as the presence of elevated levels of C-Reactive Protein and IL-6 in the blood. In men with Metabolic Syndrome, cognitive performance declined significantly over the 5-year period of the study. If these men had both Metabolic Syndrome and evidence of inflammation, cognitive performance fell even further.

Cholesterol, which is generally found in high serum levels in patients with Metabolic Syndrome, also contributes to the progression of Alzheimer's Disease[34]. Cholesterol both increases production of amyloid, as well as accelerates its accumulation in plaques in brain tissue. It also exacerbates the inflammatory response to amyloid. Moreover, studies have shown that treatment with statin drugs, used to decrease serum cholesterol, reduces risk of developing the disease. Interestingly, there are suggestions that the statins, which include drugs like Lipitor, Pravachol, and Crestor, may also reduce inflammatory processes through mechanisms other than their reduction in cholesterol synthesis.

Many questions remain to be answered. However, it appears all but certain that the hyperinsulinemia, inflammation, hyperlipidemia, hyperglycemia, hypercholesterolemia, stress, and increases in oxidative damage that are all seen in Metabolic Syndrome increase the risk and accelerate the progression of Alzheimer's Disease. To reduce the risk of Alzheimer's Disease, Metabolic Syndrome must be controlled.

REFERENCES

1. Xu, W.L. et al., Diabetes mellitus and risk of dementia in the Kungsholmen project: a 6-year follow-up study. *Neurology* 2004; 63:1181–1186.
2. Messier, C. and Teutenberg, K., The role of insulin, insulin growth factor, and insulin-degrading enzyme in brain aging and Alzheimer's Disease. *Neural Plast.* 2005; 12:311–328.
3. Luchsinger, J.A. et al., Hyperinsulinemia and risk of Alzheimer's Disease. *Neurology* 2004; 63:1187–1192.
4. Ho, L. et al., Diet-induced insulin resistance promotes amyloidosis in a transgenic mouse model of Alzheimer's Disease. *FASEB* 2004; 18:902–904.
5. Fishel, M.A. et al., Hyperinsulinemia provokes synthcronus increases in central inflammation and beta-amyloid in normal adults. *Arch. Neurol.* 2005; 62:1539–1544.
6. Margolis, R. and Altszuler, N., Insulin in the cerebrospinal fluid. *Nature* 1967; 215: 1375–1376.
7. Havrankova, J. et al., Insulin receptors are widely distributed in the central nervous system of the rat. *Nature* 1978; 272:827–829.
8. Banks, W.A., The source of cerebral insulin. *Eur. J. Pharmacol.* 2004; 490:5–12.
9. Havrankova, J. et al., Identification of insulin in rat brain. *Proc. Nat. Acad. Sci.* 1978; 75:5737–5741.
10. Marks, J.L. et al., Localization of insulin receptor RNA in rat brain by in situ hybridization. *Endocrinology* 1990; 127:3234–3236.
11. Schubert, M. et al., Insulin receptor substrate-2 deficiency impairs brain growth and promotes tau hyperphosphorylation. *J. Neurosci.* 2003; 23:7084–7092.
12. De la Monte, S.M. et al., Therapeutic rescue of neurodegeneration in experimental type 3 diabetes: relevance to Alzheimer's Disease. *J. Alzheimers Dis.* 2006; 10:89–109.
13. Shipley, M.T. and Adamek, G.D., The connections of the mouse olfactory bulb: a study using orthograde and retrograde transport of wheat germ agglutinin conjugated to horseradish peroxidase. *Brain Res. Bull.* 1984; 12:669–688.
14. Benedict, C. et al., Intranasal insulin improves memory in humans. Psychoneuroendocrinology 2004; 29:1326–1334.
15. Kern, W. et al., Improving influence of insulin on cognitive function in humans. *Neuroendocrinology* 2001; 74:270–280.
16. Tomizawa, H.H. and Halsey, Y.D., Isolation of an insulin-degrading enzyme from beef liver. *J. Biol. Chem.* 1959; 234:307–310.
17. Qiu, W.Q. and Folstein, M.F., Insulin, insulin-degrading enzyme and amyloid-beta peptide in Alzheimer's Disease: review and hypothesis. *Neurobiol. Aging* 2006; 27:190–198.
18. Kulstad, J.J. et al., Effects of chronic glucocorticoid administration on insulin-degrading enzyme and amyloid-beta peptide in the aged macaque. *J. Neuropathol. Exp. Neurol.* 2005; 64:139–146.
19. Xie, L. et al., Alzheimer's beta amyloid peptides compete for insulin binding to the insulin receptor. *J. Neurosci.* 2002; 22(RC22):1–5.
20. McLaughlin, T. et al., Metabolic changes following sibutramine-assisted weight loss in obese individuals: role of plasma free acids in the ensulin resistance of obesity. *Metabolism* 2001; 50:819–824.
21. Wada, A. et al., New twist on neuronal insulin receptor signaling in health, disease and therapeutics. *J. Pharmacol. Sci.* 2005; 99:128–143.
22. Takashima, A. et al., Activation of tau protein kinaseI/glycogen synthase kinase-3beta by amyloid beta peptide (25–35) enhances phosphorylation of tau in hippocampal neurons. *Neurosci. Res.* 1998; 31:317–323.
23. Craft, S. et al., Cerebrospinal fluid and plasma insulin levels in Alzheimer's Disease: relationship to severity of dementia and apolipoprotein E genotype. *Neurology* 1998; 50:164–168.

24. Loy, R. and Tariot, P.N., Neuroprotective properties of valproate: potential benefit for AD and tauopathies. *J. Mol. Neurosci.* 2002; 19:303–307.
25. Kunik, M.E. et al., Pharmacologic approach to management of agitation associated with dementia. *J. Clin. Psychiatry* 1994; 55(Suppl.):13–17.
26. Moretti, R. et al., Atypical neuroleptics as a treatment of agitation and anxiety in Alzheimer's Disease: risks or benefits. *Expert Rev. Neurother.* 2006; 6:705–710.
27. McGeer, P.L., Schulzer, M., and McGeer, E.G., Arthritis and anti-inflammatory agents as possible protective factors for Alzheimer's Disease: a review of 17 epidemiologic studies. *Neurology* 1996; 47:425–432.
28. Heneka, M.T. et al., Acute treatment with the PPARgamma agonist pioglitazone and ibuprofen reduces glial inflammation and Abeta1-42 levels in APPV7171 transgenic mice. *Brain* 2005; 128:1442–1453.
29. Ringheim, G.E., and Szczepanik, A.M., Brain inflammation, cholesterol, and glutamate as interconnected participants in the pathology of Alzheimer's Disease. *Curr. Pharm. Des.* 2006; 12:719–738.
30. Romanovsky, A.A. et al., Fever and hypothermia in systemic inflammation: recent discoveries and revisions. *Front. Biosci.* 10:2193–2216.
31. Pitossi, F. et al., Induction of cytokine transcripts in the central nervous system and pituitary following peripheral administration of endotoxin to mice. *J. Neurosci. Res.* 1997; 48: 287–298.
32. Sheng, J.G. et al., Lipopolysaccharide-induced neuroinflammation increases intracellular accumulation of amyloid precursor proteinand amyloid beta peptide in APPswe transgenic mice. *Neurobiol. Dis.* 2003; 14:133–145.
33. Yaffe, K. et al., The metabolic syndrome, inflammation and risk of cognitive decline. *JAMA* 2004; 18:2237–2242.
34. Kivipelto, M. and Solomon, A., Cholesterol as a risk factor for Alzheimer's Disease-epidemiological evidence. *Acta. Neurol. Scand.* 2006; 185(Suppl.):50–57.

8
METABOLIC SYNDROME, SLEEP, AND SEX

Sleep disturbances, including sleep apnea, sleep deprivation, restricted sleep, and shifted sleep schedules, can worsen Metabolic Syndrome. There is also growing evidence that Metabolic Syndrome may predispose patients to sleep disorders. Reaven has included sleep apnea among the predictable sequelae of his expanded concept of Metabolic Syndrome, that is, Insulin Resistance Syndrome. Sleep disorders are common in psychiatric illnesses, both as sequelae and as contributing factors. Thus, disturbances of sleep may be yet another link between Metabolic Syndrome and psychiatric illness.

SLEEP APNEA

Sleep apnea is simply a cessation of breathing during sleep. The common criterion to diagnose sleep apnea is multiple periods of cessation of breathing for 10 seconds or more. While this generally occurs at night, it may also happen during daytime naps. There are several types of sleep apnea. Central apnea is the least common type. It is due to abnormalities in the respiratory centers in the brain stem that prevent normal responses to changes in blood oxygen and carbon dioxide. It is most often seen in individuals with congestive heart failure, stroke or degenerative brain disorder. There is a high correlation between central sleep apnea and cardiac arrhythmia[1]. Subtle changes in the respiratory response to hypercapnia in patients with chronic respiratory disease can also result in central sleep apnea. Individuals with respiratory compromise during the day can be desensitized to hypercapnia, and this can prevent normal

Metabolic Syndrome and Psychiatric Illness: Interactions, Pathophysiology, Assessment & Treatment
Copyright © 2008 by Academic Press. All rights of reproduction in any form reserved.

response to the buildup of carbon dioxide that would ordinarily stimulate breathing at night. Thus, in some cases, central sleep apnea can be an end result of other hypoventilation disorders. A number of medications, primarily sedatives and pain relievers, can reduce the activity of the respiratory drive centers of the brain and cause decreases in respiratory activity. When alcohol is added to such medication, death can occur from the ultimate form of central apnea, that is, respiratory arrest.

The most common cause of sleep apnea is obstructive apnea. About 5 percent of adults in the United States eventually receive diagnoses of sleep apnea[2]. However, sleep apnea is woefully underdiagnosed, and as many as 31 percent of men and and 21 percent of women meet criteria that place them at high risk for this condition[3]. More than half of obese individuals meet such criteria. In obstructive sleep apnea, the airway collapses during sleep and breathing is mechanically stopped. The most obvious reason the airway is blocked is that muscles of the throat lose tone during sleep and no longer keep fatty tissue from bulging into the air passage. However, not everyone with sleep apnea is obese.

Sleep apnea is complex in that it causes two separate problems, hypoxia and loss of sleep. The cessation of breathing causes oxygen levels to plummet. In response, the brain struggles to awaken and stimulate breathing. During such an episode, a patient may partially awaken for a few seconds and quickly fall back into sleep without being aware of it. This can occur hundreds of times throughout the night. Consequently, it is not uncommon for patients with sleep apnea to report sleeping through the night but awakening in the morning more exhausted than when they went to bed.

HYPOXIA

The recurrent periods of hypoxia in sleep apnea contribute to Metabolic Syndrome. Obese men with sleep apnea have high fasting serum glucose and insulin resistance when compared to weight-matched men without sleep apnea. The degree of insulin resistance in obese men with sleep apnea correlates with the severity of their hypoxia and not their level of obesity[4]. Even in healthy men and women, hypoxia can rapidly produce resistance to insulin[5]. It is likely that the insulin resistance caused by hypoxia contributes to the hypertension, elevated triglycerides, and decreased levels of HDL that are seen in patients with sleep apnea[6]. Continuous positive pressure breathing devices can often reverse insulin resistance in these patients[7].

The chronic intermittent hypoxia of sleep apnea has been shown to adversely alter mRNA levels for adrenergic alpha2a receptors, protein mediators of leptin, and other transcripts involved in metabolic regulation in the hypothalamus[8]. Serum leptin levels also increase in hypoxia[9]. In fact, leptin stimulates respiration[10], and increases in serum leptin in hypoxia may be a compensatory response. Unfortunately, the *chronic* increases of serum leptin seen in sleep apnea do not appear to reduce appetite. They only contribute to leptin resistance. In view of leptin's ability to stimulate respiration, it would be of interest to determine if the leptin resistance often seen in obesity and Metabolic Syndrome is an independent risk factor for sleep apnea.

The hypoxia of sleep apnea also triggers inflammatory processes. Hypoxia-inducible factor-1 (HIF-1) is one of the primary mediators of cellular adaption to hypoxia. HIF-1 stimulates inflammatory responses[11], including the activation of NFk-B and other cytokines. NFk-B stimulates the release of inflammatory substances, including TNF-alpha, IL-8[12] and IL-6[13]. HIF-1 can also be activated by nonhypoxic conditions. Among such triggers of HIF-1 activation are TNF-alpha and other inflammatory cytokines that are often elevated in Metabolic Syndrome[14]. Thus, HIF-1 may be involved in a self-sustaining, upward spiral of inflammation in patients that suffer both Metabolic Syndrome and sleep apnea. Interestingly, HIF-1 is also stimulated by insulin[15]. Thus, the compensatory hyperinsulinemia of Metabolic Syndrome may participate in this circle of pathology.

Although hypoxia is responsible for many of the pathophysiological changes in sleep apnea, there is also likely to be some adverse effects arising from hypercapnia. Hypercapnia, the increase in carbon dioxide that occurs with poor respiratory ventilation, may stimulate release of rennin, vasopressin, aldosetrone, and cortisol[16]. Those changes would be expected to increase hypertension and further aggrevate Metabolic Syndrome.

Aside from hypoxia, the sleep deprivation of sleep apnea also causes adverse physiological changes. Total sleep deprivation increase serum levels of the inflammatory marker CRP in humans[17]. There have also been reports of increases in proinflammatory interleukins[18]. It should be noted that studies of sleep deprivation use models of total deprivation or curtailment of sleep in the morning. The deprivation of sleep apnea is intermittent throughout the night. Thus, there is no exact experimental model of effects of sleep loss, *per se*, in sleep apnea.

SLEEP APNEA, STRESS, AND METABOLIC SYNDROME

Studies have confirmed that sleep apnea aggravates Metabolic Syndrome; however, there is evidence that Metabolic Syndrome may increase the risk of developing sleep apnea. Although obesity can cause changes in the airway, many sufferers of sleep apnea have no obvious structural abnormalities. In fact, sleep apnea is more clearly associated with abdominal fat deposition of Metabolic Syndrome than with fatty deposits in the neck and throat that would more directly interfere with air flow[19]. TNF-alpha and IL-6 are often elevated in such patients with sleep apnea, and this has been attributed to the visceral adiposity that characterizes Metabolic Syndrome.

Hormones and cytokines related to Metabolic Syndrome may be involved in control of respiration. Leptin, for example, is a strong stimulant of respiratory drive[20]. Many with Metabolic Syndrome suffer leptin resistance, which could conceivably increase the risk of apneic events. There is evidence that certain alleles of the TNF promoter gene may predispose to obstructicve sleep apnea[21]. Treatment with etanercept, a drug that blocks the effects of TNF-alpha, reduces the number of episodes of apnea experienced during the night, as well as daytime sleepiness[22].

Although it is clear how increases in respiratory drive would lessen the risk of hypoxic episodes, it may be less obvious how changes in the neurochemical environment might decrease the risk of what may seem to be the purely mechanical obstruction of the airway. However, there are several ways by which changes in central activity could help maintain tone in the genioglossus muscle that is often responsible for airway obstruction in sleep apnea. One pharmacological tactic, which has had only marginal success, is to decrease the amount of REM (rapid eye movement) sleep and thus reduce the amount of time the body spends in the state of muscle paralysis that occurs during dreaming. It is during REM sleep that patients with obstructive sleep apnea are most vulnerable to apneic episodes[23].

Another tactic has been to enhance the activity of the hypoglossal nucleus that innervates the genioglossus muscle. It is interesting that the two medications that have so far been most successful in decreasing apneic episodes by increasing output to the genioglossus muscle are the antidepressant, mirtazapine[24] and the cholinesterase inhibitor, physostigmine[25]. The latter drug is in the class of medicines used to increase cholinergic activity to improve cognitive function in Alzheimer's Disease. It is not known what affect the various cytokines, inflammatory mediators and adipocytokines that are increased in Metabolic Syndrome might have on activity in the hypoglossal nucleus or the genioglossus muscle that it innervates.

Hypercortisolemia is not consistently observed in patients with sleep apnea; however, with time it tends to develop[26]. Some of the hypercortisolemia may be due to the stressfulness of going about one's day without proper rest. However, some of it may be due to the hypoxia of sleep apnea. Hypoxia, at least when caused by high altitude, increases serum levels of cortisol[27]. Interestingly, some individuals with hypoxia-induced hypercortisolemia at high altitudes also escape dexamethasone suppression of cortisol. This has long been seen as a characteristic of Major Depression. Hypercortisolemia would only add to the likelihood of insulin resistance and abdominal obesity. Of course, cortisol also plays an important role in regulating the sleep cycle. Thus, it is likely that hypercortisolemia of sleep apnea contributes to further sleep disturbance[28].

SMUGGLING, SNORING AND THE VELOPHARYNX

The writer and philosopher, Alan Watts, told the story of a young boy who crossed the border everyday on a bicycle. Over time, the border guard noticed that the bicycles he rode were getting bigger and more expensive. He became convinced that the boy was smuggling drugs. So, everyday he tore the boy's bicycle apart looking for the drugs, but he never found any. Each time he put the bicycle back together and reluctantly sent the boy on his way. Later he learned that the boy had been smuggling bicycles. It is possible that we focus so intently on the adverse effects of sleep apnea that we overlook the possibility that the snoring that often accompanies it may itself have adverse effects on the brain.

Although many patients have the triad of snoring, sleep apnea, and Major Depression, I am not aware of any studies suggesting that snoring might independently contribute to depression or other psychiatric disorders. I have long wondered if snoring might not effect the brain through the simple, mechanical effect of vibration. Vibration of the velopharynx, which is the primary source of snoring, takes place only 3 or 4 cm away from brainstem nuclei that supply the brain with serotonin, norepinephrine and dopamine. It is not unusual for energetic snorers to generate 80 or 90 dB of sound intensity from that tissue[29], which entails a significant amount of mechanical energy. Energy from the vibrating velopharynx is transferred into surrounding tissue. In laboratory rabbits, pressure waves from experimentally-induced snoring were discernable inside the carotid arteries[30]. From those findings it was hypothesized that in humans, the vibratory force from snoring might be sufficient to rupture plaques in the carotid arteries and send showers of emboli into the brain.

In humans, conditions such as vibration-induced Raynaud's Syndrome[31] are well known. The vibration of snoring has also been found to cause neuropathy in pharyngeal nerves that contributes to obstructive apnea[32]. In a fascinating study, human bronchial epithelial cells were cultured and subjected to vibration at a strength and frequency similar to that produced by human snorers[33]. The vibration triggered an inflammatory cascade in those cultured cells, and stimulated increases in the concentration of IL-8 in the culture medium. This was not simply mechanical disruption of cell membranes, as inhibition of synthetic pathways prevented the increase. I would very much like to know if the brain stem and brain activity is affected by the strong vibration of snoring. Some graduate students with time on their hands might want to pursue this.

SLEEP, MOOD AND GLYCOGEN SYNTHASE KINASE-3 (GSK-3)

Sleep disturbances are common among patients with mood disorders. Difficulty getting to sleep and early morning awakening are frequent complaints among sufferers of depression. Patients with BPAD, in states of mania or hypomania, may go without sleep for days Although sleep disturbance is a symptom of mood disorder, it can also contribute to the onset and progression of the illness. Thus, altering sleep patterns can be useful in the treatment of mood disorders. I have found that the single, most important thing I can do for manic patients is to get them to sleep. Conversely, sleep deprivation can be useful to jar a patient out of depression[34].

In sleep apnea, the night is punctuated with increases in autonomic activity and arousals from sleep. Although sleep deprivation can help relieve depression, the loss of sleep from apnea does not convey any antidepressant effects. Up to 20 percent of patients diagnosed with sleep apnea will also be found to have Major Depression, and vice versa[35]. It is likely that the brief episodes of sleep between episodes of arousal eliminate any potential benefits of sleep deprivation[36].

The activity of the enzyme GSK-3 is affected by hypoxia, sleep deprivation, insulin and mood stabilizers. Thus, GSK-3 may be a link between Metabolic Syndrome, sleep, and mood disorders. Recent studies using laboratory animals have revealed that hypoxia, such as occurs intermittently throughout the night in sleep apnea, increases the activity of GSK-3 in the brain[37]. One of the major effects of insulin on neurons in the brain is to reduce GSK-3 activity. Thus, physiologically, sleep apnea tends to increase insulin resistance in the brain. Like insulin, the mood stabilizers lithium and valproate, and all of the atypical antipsychotics inhibit GSK-3. Thus sleep apnea might not only increase insulin resistance by enhancing the activity of GSK-3, but also exacerbate mood disorders by the same mechanism.

GSK-3 is an enzyme with a long history in living organisms. That GSK-3 should be affected by changes in sleep may reflect the fact that it is the mammalian analogue of a chronobiologically important enzyme called SHAGGY in fruit flies[38]. SHAGGY is involved in controlling circadian rhythms in the brain of the fly. It has long been suspected that disorders in sleep and circadian rhythm play a role in the etiology of BPAD[39]. Mutations in several so-called "circadian clock genes" have been implicated as contributing to the disorder[40]. Certain variants of GSK-3 have been found to increase the likelihood that a patient with BPAD will respond to lithium[41]. The presence of that variant of GSK-3 enzyme also characterizes the subgroup of patients with BPAD whose depression improves after a night of sleep deprivation[42]. GSK-3 is clearly implicated in affective disorders and in how sleep disturbance exacerbates them.

It is not yet clear how sleep deprivation, *per se*, affects the activity of GSK-3. One possibility is mediation by endocannabinoids. Endocannabinoids are endogenous substances in the brain that have the same effects as the tetrahydrocannabinol found in cannabis. Activation of cannabinoid receptors inhibits GSK-3[43]. It is known that the substance oleamide, one of many endocannabinoids, builds up in the brains of rats that are deprived of sleep[44]. Thus it is possible that sleep deprivation increases the levels of endocannabinoids in the human brain, and these, in turn, inhibit GSK-3. Such a mechanism would be consistent with the fact that many sufferers of BPAD find relief from smoking marijuana.

SLEEP APNEA AND ALZHEIMER'S DISEASE

Sleep apnea is associated not only with affective disorders, but also with increased risk for developing Alzheimer's Disease[45]. The incidence of sleep apnea increases with age alone. Still, it is remarkable that as many as 70 percent of patients with Alzheimer's Disease may also have sleep-disordered breathing[46]. The reason for this is not entirely clear. Certainly, the damaging effects of hypoxia, stress, and lack of sleep may make the aging brain more vulnerable to the degenerative processes of Alzheimer's Disease. Hypoxia, as would occur in sleep apnea, adversely alters the expression of BACE1, an enzyme that cleaves amyloid precursor protein[47]. Aberrations of GSK-3 activity that occur with sleep disturbance may further contribute to the pathology[48].

There is growing, albeit, still controversial evidence that some individuals may be predisposed to both sleep apnea and Alzheimer's Disease due to possession of the E4 subtype of the apolipoprotein (APOE) gene. The APOE4 genotype has long been known to increase the risk of developing Alzheimer's Disease. Recent studies have shown an association between the APOE4 genotype and the likelihood of suffering sleep apnea as well[49]. If an individual has both sleep apnea and the APOE4 genotype, the risk of Alzheimer's Disease goes higher. While one cannot rid oneself of the APOE4 genotype, one can be treated for sleep apnea and reduce that risk.

INADEQUATE SLEEP AND ALTERED SLEEP SCHEDULES

Sleep problems less dramatic than apnea may also exacerbate Metabolic Syndrome. Early studies of the effects of sleep loss on metabolism revealed that several days of total sleep deprivation decreases glucose tolerance and increases insulin resistance. However, total sleep deprivation is unusual in people living average lives. What many people do suffer from is inadequate sleep. Over the last 40 years, the average American has been getting 1.5–2 hours less sleep a night. The number of young adults getting less than 7 hours of sleep a night has doubled within that time.

Two days of limiting sleep to 4 hours a night decreases glucose tolerance. After 6 days of such sleep restriction, insulin resistance rises[50]. Whereas the total amount of insulin released in response to a glucose challenge increases after 6 days of sleep restriction, the first phase of release of insulin from pancreatic beta cells decreases. Similar decreases in the first phase of insulin release is observed after lipotoxic and glucotoxic effects of poor diet damage pancreatic beta cells. It has been suggested that the changes in insulin release seen in sleep restriction may be the result of abnormal stimulation and damage of the pancreas by the sympathetic nervous system. The loss of an hour or two of sleep over prolonged periods of time increases the risk of obesity and the development of diabetes Type II[51].

Obesity is another result of a chronic lack of adequate sleep[52]. This may be due to the fact that inadequate sleep alters levels of several hormones that control appetite. For example, serum levels of the hormone ghrelin increase whereas levels of leptin decrease when sleep is restricted even a few hours[53]. Ghrelin tends to enhance appetite, whereas leptin tends to reduce it. Shift workers, that is employees working at night rather than during the usual 9 a.m. to 5 p.m. schedule, may suffer metabolic problems including obesity[54], elevated triglycerides and insulin resistance[55]. Altered sleep patterns and the stress of running contrary to normal human circadian rhythms have been offered as explanations. It is known that the most dramatic abnormalities occur just after a rapid change from day-to nighttime work, or vice versa. One interesting study evaluated metabolic changes in shift working scientists at the British Research Station in Antarctica[56]. During the dark winter months, these scientists rotate through weeklong periods of work shifted 12 hours from their usual schedules. In the day just after the shift to the new schedule, glucose tolerance was substantially reduced. Insulin resistance also rose, although this lagged one day behind the decrease in glucose tolerance. Serum levels of triglycerides and free fatty acids rose along with insulin resistance. All of those measures rapidly returned to normal after return to the usual work schedule. The data show that shifting work into the night time produces changes consistent with Metabolic Syndrome. One wonders if similar abnormalities occur in college students studying late into the night, or in business travelers suffering from jet lag.

METABOLIC SYNDROME AND SEX

A healthy sex life begins with sexual desire, or libido. For men with desire, it is necessary to achieve and maintain an erection. In women, the physiological counterparts of erection are engorgement of the clitoris and vaginal lubrication. Most would agree that orgasm is the most desirable conclusion to a sexual encounter. Unfortunately, sexual problems are quite common. In a recent British study[57], 53 percent of women and 34 percent of men reported having had a sexual problem that lasted a month or more in the previous year. Chronic sexual problems are less prevalent, with 16 percent of women and 6 percent of men reporting problems that had lasted 6 months or more. The most common sexual problems among women are lack of libido, difficulty reaching orgasm, and inadequate lubrication, in that order. Low libido is less common in men than in women, with only 5 percent of men between 18 and 59 reporting a lack of sexual desire[58]. However, about 20 percent of men over 40 complain of episodes of erectile dysfunction, and this increases with age[59].

Psychiatric illnesses themselves can cause or aggrevate problems of sexual function. Certainly, some of the problems can arise from medications used to treat the illness[60]. However, up to 65 percent of patients being treated for depression experience sexual dysfunction even before the start of treatment with antidepressants[61]. Good sexual function should be the goal in the treatment of every patient, particularly those wounded by anxiety and affective disorders. As much as possible should be done to optimize the likelihood of maintaining or recovering sexual function. Sexual activity requires adequate blood flow, healthy endothelial tissue, and normal activity of the endocrine and nervous systems. Metabolic Syndrome affects hormones, the cardiovascular system, and the brain. Consequently, Metabolic Syndrome can have significant effects on human sexuality, and mainteneance or restoration of sexual function may depend upon successfully addressing this condition.

Low testosterone and Metabolic Syndrome go together. Young men with Metabolic Syndrome have lower serum testosterone levels than men their age without the condition[62]. The expected, age-related decline in testosterone is made worse by the presence of Metabolic Syndrome[63]. Men with Klinefelter's Syndrome, who invariably have low serum testosterone levels, also tend to be insulin resistant[64]. Elderly men with testosterone levels in the higher ranges of normal have better sensitivity to insulin than men with lower testosterone[65]. What is not clear is whether low testosterone contributes to Metabolic Syndrome or vice versa. In fact, there is evidence for effects going both ways.

In men with resistance to insulin, the Leydig cells of the testes produce lower than normal amounts of testosterone[66]. Moreover, when men with the abdominal obesity that characterizes Metabolic Syndrome lose extra weight, their testosterone levels often return to normal[67]. Thus, certain aspects of Metabolic Syndrome may lower serum testosterone levels. On the other hand, men with low testosterone are up to four times more likely to develop Metabolic Syndrome than men with normal testosterone levels[68]. This would suggest that normal testosterone levels help men avoid Metabolic Syndrome. Unfortunately, most studies evaluating the effects of testosterone supplementation on insulin sensitivity have been equivocal. While low tesoterone may contribute to Metabolic Syndrome, it is probably not a major factor in its development.

In both men and women, testosterone is the hormone primarily responsible for sex drive. Because of decreases in serum testosterone levels seen in men with Metabolic Syndrome, it is not surprising that a large percentage of men with this condition report decreases in libido and reduced frequency of sexual intercourse[69]. There is also strong evidence that Metabolic Syndrome increases the likelihood of erectile dysfunction[70]. Men with erectile dysfunction have nearly twice the incidence of Metabolic Syndrome than do age-matched men without erectile dysfunction. After men change their diets and improve their signs of Metabolic Syndrome, their sexual function also improves. The so-called Mediterranean Diet , which is high in olive oil, whole grains, fruits and nuts, can be very helpful in preventing and reversing Metabolic Syndrome. Men who consume a Mediterranean style diet tend to have less incidence of erectile dysfunction. Moreover, men who lose weight and reverse their signs of Metabolic Syndrome by following a Mediterranean Diet tend to experience reversal of erectile dysfunction they have suffered.

The most obvious relationship between Metabolic Syndrome and erectile dysfunction lies in changes in the health of the lining of arteries, the endothelium, that occur as Metabolic Syndrome progresses. High serum levels of CRP and pro-inflammatory cytokines, such as IL-6 and TNF-alpha, are found both in Metabolic Syndrome and in men with erectile dysfunction[71]. These inflammatory substances contribute to endothelial damage. Damage to the endothelium prevents production of nitric oxide[72], which causes relaxation of the arteries to bring more blood into the penis. Many authors have noted that deficits in blood flow to the penis that cause erectile dysfunction often mirror poor blood flow through the coronary arteries of the heart, and serve as a warning sign of coronary artery disease. The progression from Metabolic Syndrome to outright diabetes Type II is a poor prognostic sign for men with erectile dysfunction, as the evidence shows that men with diabetes do not respond well to drugs used to treat erectile dysfunction such as Viagra.

In view of the high incidence of sleep apnea in patients with Metabolic Syndrome, it is also worth noting that erectile dysfunction is a common problem in men with sleep apnea[73]. Nearly 80 percent of men with significant sleep apnea report having erectile dysfunction as well. There is a significant correlation between the severity of sleep apnea and that of the erectile dysfunction. Moreover, treatment with CPAP can reverse the erectile dysfunction in up to 75 percent of these men[74]. The mechanism by which sleep apnea causes erectile function is not entirely clear. It could be due in part to decrease in endothelial function[75]. Decreases in activity of the pituitary-gonadal axis may also be involved. Men with sleep apnea have decreased serum levels of both luteinizing hormone and testosterone[76]. This might help explain the decrease in libido that is also reported in men with sleep apnea. Thus, sleep apnea may be one more way in which sexual performance can be compromised in Metabolic Syndrome.

Women with Metabolic Syndrome also complain of diminished pleasure and satisfaction in sexual relations[77]. Women with Metabolic Syndrome have sexual desire similar to that of women without Metabolic Syndrome. However, they complain of problems with arousal, vaginal lubrication and achieving orgasm. As with men, women with Metabolic Syndrome have elevated serum levels of inflammatory substances that reduce synthesis of nitric oxide in the endothelium. In women, nitric oxide increases blood flow to the clitoris. Clitoral engorgement with blood flow is thought to be a necessary component of full female sexual response. The use of drugs such as Viagra that increase blood flow to the penis is helpful in men with erectile dysfunction. However, this class of medicines has not consistently been found to be useful in women with sexual dysfunction. In view of the fact that these medications act largely by enhancing the effects of nitric oxide, they might be of more value in women with Metabolic Syndrome who also suffer inability to achieve orgasm.

It is possible that some of the adipocytokines that are elevated in the blood of patients with Metabolic Syndrome have adverse effects on sexual behavior. There is evidence from animal and human studies that leptin is important in sexual behavior and reproduction. In Metabolic Syndrome, serum leptin levels and sensitivity to the hormone are often disturbed. As is the case with insulin in Metabolic Syndrome, the body may suffer both from resistance to leptin as well as to effects of compensatory hyperleptinemia. Leptin is necessary for normal reproduction in humans, as it stimulates the brain to release hormones that cause the release of estrogen and progesterone from the ovaries. Women who lose too much weight stop having normal menstrual periods. This is partially due to the depletion of fat cells that produce and release leptin. Because fat is decreased in the eating disorder anorexia nervosa, it is not surprising that leptin levels are extremely low in the mostly young women who have this illness[78]. It is thought that decreased levels of leptin are largely responsible for the discontinuation of menstrual cycling in women with anorexia nervosa[79]. Obese women also have menstrual irregularities and problems with fertility. It is possible that hyperleptinemia and leptin resistance are at least partially responsible[80].

It has recently been discovered that deletion of one component of the intracellular second messenger system of insulin, the insulin receptor substrate-2, results in both metabolic disturbance and infertility in female mice[81]. These mice had increased levels of food intake and obesity. However, they also had low serum levels of sex steroids, luteinzing hormone and prolactin. Interestingly, they also had high serum levels of leptin. Thus, an intact insulin receptor signaling system may be required for normal levels of sex hormones and fertility in female mammals.

Administration of leptin directly into the brains of diabetic male rats, increases the number of ejaculations in comparison with rats receiving only saline[82]. Leptin also tended to shorten the amount of time it took for males to become interested in mounting females again after they ejaculated. Generally, such effects are interpreted as enhancement of motivation for sexual activity. It is not known if leptin has any such effect in humans.

Men with Metabolic Syndrome and obesity have an unexpectedly high rate of premature ejaculation. The mechanism by which this might occur is not clear. There are reports that a large percentage of men with diabetes experience premature ejaculation. Moreover, the risk of this problem is greatest when glycemic control is poor[83]. Men with elevated glycosylated hemoglobin levels, which serve as evidence of persistently high serum glucose, are nearly ten times as likely to have premature ejaculation. Thus, it is possible that oxidative damage or some other ill effect of hyperglycemia is responsible. However, hyperleptinemia is also associated with premature ejaculation in men[84]. Knowing the relationships between obesity, Metabolic Syndrome and leptin, it is tempting to suggest that the premature ejaculation seen in men with Metabolic Syndrome is due to hyperleptinemia. As an aside, I must also note my fascination with the fact that spermatozoa are among the few cells in the human body that secrete both leptin and insulin[85]. Defects in the production of these substances by spermatozoa might be a cause of infertility in some men.

REFERENCES

1. Lanfranchi, P.A. et al., Central sleep apnea in left ventricular dysfunction: prevalence and implications for arrhythmic risk. *Circulation* 2003; 107:727–732.
2. Young T., Peppard, P.E., and Gottlieb, D.J., Epidemiology of obstructive sleep apnea: a population health perspective. *Am. J. Respir. Crit. Care Med.* 2002; 165:1217–1239.
3. Hiestand, D.M. et al., Prevalence of symptoms and risk of sleep apnea in the US population: results from the national sleep foundation sleep in America 2005 poll. *Chest* 2006; 130: 780–786.
4. Punjabi, N.M. et al., Sleep-disordered breathing and insulin resistance in middle-aged and overweight men. *Am. J. Respir. Crit. Care Med.* 2002; 165:677–682.
5. Punjabi, N.M. and Polotsky, V.Y., Disorders of glucose metabolism in sleep apnea. *J. Appl. Physiol.* 2005; 99:1998–2007.
6. Coughlin, S.R., et al. Obstructive sleep apnoea is independently associated with an increased prevalence of metabolic syndrome. *Eur. Heart J.* 2004; 25:735–741.
7. Harsch, I.A. et al., Continuous positive airway pressure treatment rapidly improves insulin sensitivity in patients with obstructive sleep apnea syndrome. *Am. J. Respir. Crit. Care Med.* 2004; 169:156–162.
8. Volgin, D.V. and Kubin, L., Chronic intermittent hypoxia alters hypothalamic transcription of genes involved in metabolic regulation. *Auton. Neurosci.* 2006; 126:93–99.
9. Yingzhong, Y. et al., Regulation of body weight by leptin, with special reference to hypoxia-induced regulation. *Intern. Med.* 2006; 45:941–946.
10. O'Donnel, C.P. et al., Leptin prevents respiratory depression in obesity. *Am. J. Respir. Crit. Care Med.* 1999; 159:1477–1484.
11. Hierholzer, C. and Billiar, T.R., Molecular mechanisms in the early phase of hemorrhagic shock. *Langenbecks Arch. Surg.* 2001; 386:302–308.
12. Ryan, S., Taylor, C.T., and McNicholas, W.T., Predictors of elevated nuclear factor-kappaB-dependent genes in obstructive sleep apnea syndrome. *Am. J. Respir. Crit. Care Med.* 2006; 174:824-830.
13. Vgontzas, A.N. et al., Elevation of plasma cytokines in disorders of excessive daytime sleepiness: role of sleep disturbance and obesity. *J. Clin. Endocrinol. Metab.* 1997; 82:1313–1316.
14. Albina, J.E. et al., HIF-1 expression in healing wounds: HIF-1alpha induction in primary inflammatory cells by TNF-alpha. *Am. J. Physiol. Cell. Physiol.* 2001; 281:C1971–1977.
15. Carnesecchi, S. et al., Insulin-induced vascular endothelial growth factor expression is mediated by the NADPH oxidase NOX3. *Exp. Cell. Res.* 2006; 312:3413–3424.
16. Saaresranta, T. and Polo O., Sleep-disordered breathing and hormones. *Eur. Respir. J.* 2003; 22:161–172.
17. Meier-Ewert, H.K. et al., Sleep loss contributes to elevations in inflammatory response. *J. Am. Coll. Cardiol.* 2004; 43:678–683.
18. Moldofsky, H. et al., Effects of sleep deprivation on human immune functions. *FASEB* 1989; 3:1972–1977.
19. Vgontzas, A.N., Bixler, E.O., and Chrousos, G.P., Metabolic disturbances in obesity versus sleep apnoea: the importance of visceral obesity and insulin resistance. *J. Intern. Med.* 2003; 254:32–44.

20. Campo, A. et al., Hyperleptinemia, respiratory drive and hypercapnic response in obese patients. *Eur. Respir. J.* 2007 (in print).

21. Liu, H.G. et al., [The relationship between tumor necrosis factor-alpha gene promoter polymorphism and obstructive sleep apnea-hypopnea syndrome] *Zhonghua Jie He He Hu Xi Za Zhi* 2006; 29:596–599.

22. Vgontzas, A.N. et al., Marked decrease in sleepiness in patients with sleep apnea by etanercept, a tumor necrosis factor-alpha antagonist. *J. Clin. Endocrinol. Metab.* 2004; 89:4409–4413.

23. Veasey, S.C. et al., Medical therapy for obstructive sleep apnea: a review by the Medical Therapy for Obstructive Sleep Apnea Task Force of the Standards of Practice Committee of the American Academy of Sleep Medicine. *Sleep* 2006; 29:1036–1044.

24. Carley, D.W. et al., Efficacy of mirtazapine in obstructive sleep apnea syndrome. *Sleep* 2007; 30:35–41.

25. Hedner, J. et al., Reduction of sleep-disordered breathing after physostigmine. *Am. J. Respir. Crit. Care Med.* 2003; 168:1246–1251.

26. Buckley, T.M. and Schatzberg, A.F., On the interactions of the hypothalamic-pituitary-adrenal (HPA) axis and sleep: normal HPA axis activity and circadian rhythm, exemplary sleep disorders. *J. Clin. Endocrinol. Metab.* 2005; 90:3106–3114.

27. Martignoni, E. et al., The effects of physical exercise at high altitude on adrenocortical function in humans. *Funct. Neurol.* 1997; 12:339–344.

28. Fehm, H.L. et al., Influences of corticosteroids, dexamethasone and hydrocortisone on sleep in humans. *Neuropsychobiology* 1986; 16:198–204.

29. Hoffstein, V., et al., Does snoring contribute to presbycusis? *Am. J. Respir. Crit. Care Med.* 1999; 159:1351–1354.

30. Amatoury, J., et al., Snoring related-related energy transmission to the carotid artery in rabbits. *J. Appl. Physiol.* 2006; 100:1547–1553.

31. Herrick, A.L., Pathogenesis of Raynaud's phenomenon. *Rheumatology (Oxford)* 2005; 44:587–596.

32. Friberg, D., Heavy Snorer's Disease: A Progressive Local Neuropathy. *Acta Oto-Laryngologica* 1999; 119:925–933.

33. Puig, F. et al., Vibration enhances interleukin-8 release in a cell model of snoring-induced airway inflammation. *Sleep* 2005; 28:1312–1316.

34. Giedke, H. et al., Direct comparison of total sleep deprivation and late partial sleep deprivation in the treatment of major depression. *J. Affect. Disord.* 2003; 76:85–93.

35. Schroder, C.M. and O'Hara, R., Depression and obstructive sleep apnea. Depression and Obstructive Sleep Apnea (OSA). *Ann. Gen. Psychiatry* 2005; 4:13–33.

36. Wirz-Justice, A. and Van den Hoofdakker, R.H., Sleep deprivation: What do we know? Where do we go? Biol. Psychiatry 1999; 46:445–453.

37. Goldbart, A. et al., Intermittent hypoxia induces time-dependent changes in the protein kinase B signaling pathway in the hippocampal CA1 region of the rat. *Neurobiol. Dis.* 2003; 14:440–446.

38. Martinek, S. et al., A role for the segment polarity gene shaggy/GSK-3 in the Drosophila circadian clock. *Cell* 2001; 105:769–779.

39. Boivin, D.B., Influence of sleep-wake and circadian rhythm disturbances in psychiatric disorders. *J. Psychiatry Neurosci.* 2000; 25:446–458.

40. Nievergelt, C.M. et al., Suggestive evidence for association of the circadian genes PERIOD3 and ARNTL with bipolar disorder. *Am. J. Med. Genet. B Neuropsychiatr. Genet.* 2006; 141:234–241.

41. Benedetti, F. et al., Long-term response to lithium salts in bipolar illness is influenced by the glycogen synthase kinase 3-beta-50 T/C SNP. *Neurosci. Lett.* 2005; 376:51–55.

42. Benedetti, F. et al., A glycogen synthase kinase 3-beta promoter gene single nucleotide polymorphism is associated with age at onset and response to total sleep deprivation in bipolar depression. *Neurosci. Lett.* 2004; 368:123–126.

43. Gómez del Pulgar, T., Velasco, G., and Guzmán, M., The CB1 cannabinoid receptor is coupled to the activation of protein kinase B/Akt. *Biochem. J.* 2000; 347(Pt 2):369–373.

44. Mendelson, W.B. and Basile, A.S., The hypnotic actions of oleamide are blocked by a cannabinoid receptor antagonist. *Neuroreport* 1999; 10:3237–3239.

45. Bliwise, D.L., Sleep apnea, APOE4 and Alzheimer's Disease 20 years and counting? *J. Psychosom. Res.* 2002; 53:539–546.

46. Chong, M.S., Continuous positive airway pressure reduces subjective daytime sleepiness in patients with mild to moderate Alzheimer's Disease with sleep disordered breathing. *J. Am. Geriatr. Soc.* 2006; 54:777–781.

47. Sun, X. et al., Hypoxia facilitates Alzheimer's Disease pathogenesis by up-regulating BACE1 gene expression. *Proc. Natl. Acad. Sci. U.S.A* 2006; 103:18727–18732.

48. Forde, J.E. and Dale, T.C., Glycogen synthase kinase 3: a key regulator of cellular fate. *Cell. Mol. Life. Sci.* May 29, 2007 (Epublished ahead of print).

49. Kadotani, H. et al., Association between apolipoprotein E4 and sleep-disordered breathing in adults. *JAMA* 2001; 285:2888–2890.

50. Spiegel, K. et al., Sleep loss: a novel risk factor for insulin resistance and Type 2 diabetes. *J. Appl. Physiol.* 2005; 99:2008–2019.

51. Copinschi, G., Metabolic and endocrine effects of sleep deprivation. *Essent. Psychopharmacol.* 2005; 6:341–347.

52. Taheri, S., The interactions between sleep, metabolism and obesity. *Int. J. Sleep Wakefulness* 2007; 1:20–29.

53. Spiegel, K. et al., Brief communication: sleep curtailment in healthy young men is associated with decreased leptin levels, elevated ghrelin levels, and increased hunger and appetite. *Ann. Intern. Med.* 2004; 141:846–850.

54. Di Lorenzo, L. et al., Effect of shift work on body mass index: results of a study performed in 319 glucose-tolerant men working in a Southern Italian industry. *Int. J. Obes. Relat. Metab. Disord.* 2003; 27:1353–1358.

55. Scheen, A.J., [Clinical study of the month. Does chronic sleep deprivation predispose to metabolic syndrome?] *Rev. Med. Liege.* 1999; 54:898–900.

56. Lund, J. et al., Postprandial hormone and metabolic responses amongst shift workers in Antarctica. *J. Endocrinol.* 2001; 171:557–564.

57. Mercer, C.H. et al., Sexual function problems and help seeking behavior in Britain: national probability sample survey. *BMJ* 2003; 327:426–427.

58. Laumann, E.O., Palk, A., and Rosen, R.C., Sexual dysfunction in the United States: prevelance and predictors. *JAMA* 1999; 28:537–544.

59. Wylie, K.R., Machin, A., *Prim. Psychiatry* 2007; 14:65–71.

60. Balon, R., Depression, antidepressants and human sexuality. *Prim. Psychiatry* 2007; 14:47–50.

61. Bonierbale, M., Lancon, C., and Tignol, J., The ELIXER study: evaluation of sexual dysfunction in 4557 depressed patients in France. *Curr. Med. Res. Opin,* 2003; 19:114–124.

62. Robeva, R. et al., Low testosterone levels and unimpaired melatonin secretion in young males with metabolic syndrome. *Andrologia* 2006; 38:216–220.

63. Kaplan, S.A., Meehan, A.G., and Shah, A., The age related decrease in testosterone is significantly exacerbated in obese men with the metabolic syndrome. What are the implications for the relatively high incidence of erectile dysfunction observed in these men? *J. Urol.* 2006; 176(4 Pt 1):1524–1527.

64. Pei, D. et al., Insulin resistance in patients with Klinefelter's syndrome and idiopathic gonadotropin deficiency. *J. Formos. Med. Assoc.* 1998; 97:534–540.

65. Muller, M. et al., Endogenous sex hormones and metabolic syndrome in aging men. *J. Clin. Endocrinol. Metab.* 2005; 90:2618–2623.

66. Pitteloud, N. et al., Increasing insulin resistance is associated with a decrease in Leydig cell testosterone secretion in men. *J. Clin. Endocrinol. Metab.* 2005; 90:2636–2641.

67. Niskanen, L. et al., Changes in sex hormone-binding globulin and testosterone during weight loss and weight maintenance in abdominally obese men with the metabolic syndrome. *Diabetes Obes. Metab.* 2004; 6:208–215.

68. Kupelian, V. et al., Low sex hormone-binding globulin, total testosterone, and symptomatic androgen deficiency are associated with development of the metabolic syndrome in nonobese men. *J. Clin. Endocrinol. Metab.* 2006; 91:843–850.

69. Corona, G. et al., Psychobiologic correlates of the metabolic syndrome and associated sexual dysfunction. *Eur. Urol.* 2006; 50:595–604.

70. Esposito, K. et al. Mediterranean diet improves erectile function in subjects with the metabolic syndrome. *Int. J. Impot. Res.* 2006; 18:405–410.

71. Vlachopoulos, C. et al., Unfavourable endothelial and inflammatory state in erectile dysfunction patients with or without coronary artery disease. *Eur. Heart J.* 2006; 27:2640–2648.

72. Huang, A.L. and Vita, J.A., Effects of systemic inflammation on endothelium-dependent vasodilation. *Trends Cardiovasc. Med.* 2006; 16:15–20.

73. Teloken, P.E., et al., Defining association between sleep apnea syndrome and erectile dysfunction. *Urology* 2006; 67:1033–1037.

74. Goncalves, M.A. et al., Erectile dysfunction, obstructive sleep apnea syndrome and nasal CPAP treatment. *Sleep Med.* 2005; 6:333–339.

75. Lattimore, J.L. et al., Treatment of obstructive sleep apnea leads to improved microvascular endothelial function in the systemic circulation. *Thorax* 2006; 61:459–460.

76. Luboshitzky, R. et al., Decreased pituitary-gonadal secretion in men with obstructive sleep apnea. *J. Clin. Endocrinol. Metab.* 2002; 87:3394–3398.

77. Esposito, K. et al., The metabolic syndrome: a cause of sexual dysfunction in women. *Int. J. Impot. Res.* 2005; 17:224–226.

78. Krassas, G.E., Endocrine abnormalities in Anorexia Nervosa. *Pediatr. Endocrinol. Rev.* 2003; 1:46–54.

79. Holtkamp, K. et al., Reproductive function during weight gain in anorexia nervosa. Leptin represents a metabolic gate to gonadotropin secretion. *J. Neural Transm.* 2003; 110:427–435.

80. Bellver, J. et al., Obesity and assisted reproductive technology outcomes. *Reprod. Biomed. Online* 2006; 12:562–568.

81. Burks D.J. et al., IRS-2 pathways integrate female reproduction and energy homeostasis. *Nature* 2000; 407:377–382.

82. Saito, T.R. et al., Effects of i.c.v. administration of leptin on copulatory and ingestive behavior in STZ-induced diabetic male rats. *Exp. Anim.* 2004; 53:445–451.

83. El-Sakka, A.I. Premature ejaculation in non-insulin-dependent diabetic patients. *Int. J. Androl.* 2003; 26:329–334.

84. Atmaca, M. et al., Serum leptin levels in patients with premature ejaculation. *Arch. Androl.* 2002; 48:345–350.

85. Andò, S. and Aquila, S., Arguments raised by the recent discovery that insulin and leptin are expressed in and secreted by human ejaculated spermatozoa. *Mol. Cell. Endocrinol.* 2005; 245:1–6.

9
DIETS FOR WEIGHT LOSS AND METABOLIC SYNDROME

No doctor or dietician would take exception to the fact that when the number of calories a person consumes falls below the number of calories their body burns, that person will lose weight. The reverse is also true. If a person consumes more calories than their body uses, they will gain weight. Given these facts, it would seem that the approach to losing weight and maintaining a healthy body mass index would be simple and straightforward. Hah! Unfortunately, the science of diet and weight loss continues to be very tricky business.

Diets are complicated for a number of reasons. Although reduction of the intake of calories is the primary cause of weight loss, the source of those calories must be carefully considered. Blood sugar, insulin levels, cholesterol, triglycerides and other physiological measures can all be influenced by whether the calories come from protein, carbohydrate or fat. Thus, the effectiveness and tolerability of diet must be weighed against effects on cardiovascular and other aspects of health directly related to Metabolic Syndrome.

Whether calories come from protein, carbohydrate or fat can also make a difference in how satisfying those calories are and, in turn, how many of those calories a person wants to consume. There are even differences in how efficiently the body burns calories from different sources. Although all calories may be equal in the test tube, in the human body one calorie may be worth more than another in the amount of energy that can actually be extracted from it. There are even indications that some nutrients can stimulate the burning of calories. These facts prompt the question as to whether there might be diets that, beyond merely reducing calorie intake, can also help people lose weight more quickly and comfortably.

There are basically three types of diets for weight loss. One is simply reducing calorie intake, that is, eating what you always eat but less of it.

Metabolic Syndrome and Psychiatric Illness: Interactions, Pathophysiology, Assessment & Treatment
Copyright © 2008 by Academic Press. All rights of reproduction in any form reserved.

Aside from the benefits of weight loss, this restriction of calories is no more or less healthy than the style of eating that preceded it. It is the safety and effectiveness of the other two types of diets that has been a source of controversy among physicians and nutritional scientists. The first of these diets is the low-fat and high-carbohydrate diet. The other is the increasingly popular low-carbohydrate, high-protein and high-fat diet.

LOW-FAT, HIGH-CARBOHYDRATE DIETS

At first, the logic of low-fat, high-carbohydrate diets seemed simple and flawless. Fat is very high in calories. Ounce for ounce, fat has twice the calories of carbohydrate and protein. Foods high in fat are also often rich in cholesterol, which can contribute to heart disease. The most reasonable conclusion seemed to be that if you reduce your fat intake and rely more on carbohydrates for energy, you will lose weight and improve the health of your heart. This conclusion led to the development of low-fat, high-carbohydrate diets for weight loss and improvement of cardiovascular health.

In 2001, the American Heart Association published its Therapeutic Lifestyle Changes (TLC) diet, which replaced the previous Step I and Step II diets. In the TLC diet, it is recommended that fat account for no more than 30 percent of a person's daily intake of calories and saturated fat make up no more than 10 percent of the total calories. It was further recommended that 50–60 percent of calories come from carbohydrate and the remaining 10–20 percent of calories from protein. Diets even more restrictive of fats have also been devised, including the Ornish and Pritikin diets. The Ornish diet restricts fat intake to only 10 percent of calories, with 70 percent coming from carbohydrate and 20 percent from protein. Unfortunately, low-fat diets are not only unpalatable and difficult to maintain, but it turns out that for most people this type of diet gives little, if any, protection from cardiovascular disease[1].

Although the relative ineffectiveness of low-fat, high-carbohydrate diets in reducing cardiovascular risk may be nonintuituve, it is easily explained by the pathophysiology of Metabolic Syndrome. Some people exhibit increases in levels of triglycerides and VLDL, as well as decreases in HDL, when they consume large quantities of carbohydrate. This carbohydrate-induced hyperlipidemia is largely a matter of genetic predisposition to insulin resistance. Whereas some people may benefit from a reduced calorie, low-fat, high-carbohydrate diet, people predisposed to insulin resistance will not. In fact, they are likely to exacerbate Metabolic Syndrome and increase their risk of cardiovascular disease.

Rather unexpectedly, it turns out that for many people there are advantages to diets that reduce carbohydrate and replace those calories with fat and protein. These diets not only enhance weight loss, but they also reduce the manifestations of Metabolic Syndrome and thereby reduce cardiovascular risk. It is likely that they will also lower risk for psychiatric illnesses, Alzheimer's dementia, sexual dysfunction, sleep apnea and other expected sequelae of the syndrome.

A FEW WORDS ABOUT PROTEIN

Many diets have been devised for weight loss in patients at risk for Metabolic Syndrome. For the most part, these diets are based on reduction of carbohydrate intake and replacement of the lost calories with those from fat and protein. The most extreme example of this type of diet is the controversial Atkin's Diet. However, there are also newer and more conservative proposals for diets that replace carbohydrate with "good fat", and thus avoid some of the concerns about consumption of too much saturated fat. There are compelling reasons to believe that this type of diet has some advantages. To better explain the merits of high-fat, high-protein diets, I will briefly review the role of protein in nutrition and physiology.

Twenty different amino acids make up the proteins in the human body. Most of those amino acids can be made in the body; however, nine cannot be synthesized and are referred to as "essential". They include valine, tryptophan, threonine, phenylalanine, methionine, lysine, leucine, and isoleucine. The essential amino acids must be obtained in the diet.

Proteins from food that match the types and proportions of amino acids in the human body are said to have high biological value. The proteins with the highest biological value are in eggs. Those in milk and meat follow closely behind. Proteins from plant sources usually do not have high biological value. However, if foods of plant origin are eaten together in proper proportions, such as combinations of beans and grains, adequate intake of the essential amino acids can be obtained.

Although amino acids can be assembled back into proteins or transformed into neurotransmitters, they can also be used for fuel. It is the use of protein and its constituent amino acids as a source of fuel that has the most significant effect on the metabolism of glucose and fat, the action of insulin, control of weight, and Metabolic Syndrome.

Although insulin is seen primarily in the context of its effects on glucose and fat metabolism, it also has important effects on protein metabolism. Insulin enhances the uptake of amino acids into cells. It enhances protein synthesis and inhibits the breakdown of protein back into its constituent amino acids. By causing amino acids to be taken up into cells and utilized to make protein, insulin minimizes the likelihood that they will be transported to the liver and turned into glucose.

Ingestion of protein stimulates release of insulin. This effect of protein is thought to be due primarily to increases in circulating levels of the amino acids leucine and arginine. However, most of the other amino acids also contribute to the stimulation of insulin release. The effect of amino acids on insulin release can be as strong as the stimulating effect of glucose. The release of insulin is even greater when protein and carbohydrates are consumed together.

All of the amino acids, with the exception of leucine, can be converted into glucose. Those amino acids are "glucogenic". Leucine, as well as several of the other amino acids, can be metabolised into ketones. Those amino acids are referred to as "ketogenic". Because most of the amino acids in protein are glucogenic, the primary fuel product of protein intake is glucose. This fact has been the source of many misunderstandings. Some authors of diet and nutrition books have warned their readers that the conversion of amino acids to glucose causes high protein diets to be little different from high carbohydrate diets. However, the scientific litereature shows these concerns to be unfounded.

An elegant study was performed in 1936 by Drs. Jerome Conn and Louis Newburgh[2]. This is the same Jerome Conn with whom Dr. Gerald Reaven worked for several years prior to his permanent position at Stanford University. They set out to test the prevailing notion that high protein diets were dangerous for individuals with diabetes. At that time it was believed that the conversion of the amino acids in the protein into glucose could worsen hyperglycemia. They compared the effects of isocaloric amounts of protein and carbohydrate on serum glucose levels. The serum glucose levels in the subjects receiving carbohydrate shot up within 45 minutes to an hour and, after another hour or so, often dipped below starting level. The glucose levels of those that received protein held rock steady the entire time. They concluded that the slow liberation of glucose from protein could be very helpful, not only to the diabetic patient but also to those that suffer postprandial hypoglycemia. Modern studies have continued to confirm the results of Conn and Newburgh. Protein meals do not cause release of bursts of glucose into the blood. Rather, a slow conversion of some of the protein into glucose, and the serum glucose levels remain steady[3].

An unavoidable side effect of increasing protein intake is a subsequent increase in serum urea. Patients with hepatitis or cirrhosis do not tolerate high protein intake because the liver is not able to keep up with turning the ammonia into urea. In fact, such people cannot even tolerate the usual production of ammonia by bacteria in the intestinal tract. Certainly, high protein diets should be avoided by such people[4]. Individuals in renal failure should also avoid high protein[5].

Meat also contains significant amounts of RNA and DNA that contain purines. Purines are transformed into uric acid. In certain individuals levels of uric acid can build up. This can occur when production of uric acid is too high or excretion from the body is too low. The latter may happen due to kidney disease or hereditary conditions. The result is the exquisitely painful condition known as gout. People with gout should not be on high protein diets if the source of the protein is meat, particularly high purine-containing meat such as liver or kidney[6].

DIET, APPETITE, AND SATIETY

Many diets make claims of having almost magical ability to melt pounds away. However, there is no magic in dieting. If a diet is going to have any special ability to bring about weight loss, aside from its mere restriction of calorie intake, it can only come about in one of two ways. Such diets either increase the satiating effects of each meal and reduce appetite for the next meal, or they increase the rate at which calories are burned.

Studies have shown that meals high in protein produce greater satiety than isocaloric meals high in fat or carbohydrate[7]. People are simply not as hungry after eating a meal high in protein. However, the satiating effect of protein tends to diminish with time, and it can disappear after only a few weeks on a high-protein diet[8]. The advantage of high-protein diets in producing weight loss also tends to diminish. High-protein diets seem to be better for weight loss within the first months of the diet.

Not only are high protein meals more satiating, they also lower appetite for meals later in the day. When compared with subjects eating high-fat or high-carbohydrate meals, those eating high amounts of protein eat less in their next meal. This difference is particularly striking when high-protein meals are compared with meals rich in high glycemic index carbohydrates[9]. Reactive hypoglycemia, which can occur several hours after intake of quickly digested carbohydrates, can create feelings of hunger in some individuals. Thus, aside from any effects of the protein itself, high-protein diets may indirectly cut hunger between meals by minimizing the consumption of such carbohydrates.

Because fat and carbohydrate are far better sorces of energy, it is unlikely that the ability of protein to increase satiety and reduce hunger is due simply to its caloric content. It is more likely that something in protein, perhaps one or more amino acids, serves as a special signal of satiety. Leucine is a likely candidate for this role. This may occur through direct action in the brain, or indirectly through stimulation of some intermediary substance that later acts in the brain to reduce appetite.

There are many substances released into the blood after food is eaten that affect hunger and satiety. Protein is more effective than carbohydrate in suppressing the appetite stimulating hormone ghrelin. Protein also increases levels of the appetite suppressing hormones cholecystokinin and glucagon-like peptide-1[10].

Replacing carbohydrate with protein and fat also causes ketones to appear in the blood. The presence of ketones in this context is not an abnormality, as it might be seen to be in a patient with diabetes. The use of ketones for fuel is a perfectly natural and healthy form of energy metabolism. In fact, ketones are the favored fuel of the heart, and they are readily burned in muscle cells. Although the brain prefers glucose, it is also able to use ketones. This ability of the brain to use ketones for fuel increases during periods of fasting or starvation.

It has long been suspected that increased levels of ketones in the blood can decrease appetite. However, there is astonishingly little scientific evidence for this. There is even evidence to the contrary[11]. Nonetheless, there are some fascinating physiological studies showing that ketones may have some effects on cellular metabolism that are similar to those of insulin[12]. During periods of starvation insulin levels drop precipitously, whereas ketone bodies increase in concentration. Thus, in the absence of insulin, there may be a physiological basis for ketones to mimic some of the metabolic control mechanisms that insulin normally provides. Both ketones and insulin cause translocation of GLUT4 glucose transporters to cell membranes, increase availability of substrates early in the Kreb cycle, increase mitochondrial acetyl-CoA concentration, and increase the efficency of heart muscle. Perhaps ketones also mimic some of the satiety effects that insulin is known to produce in the hypothalamus.

Because the presence of ketones is a marker for the lack of carbohydrate, suppression of appetite may not be from the ketones, but rather from ending the roller-coaster effect of consuming too much high glycemic index carbohydrate. Low-carbohydrate diets that generate ketones also tend to include higher than usual intake of protein, which maybe a major factor in appetite suppression. Part of the reduction of appetite on high-fat, high-protein diets could be due to increases in serum leucine. Curiously, high amounts of fat in the diet can also increase serum levels of leucine. In rats fed a diet of 90 percent fat, the concentration of leucine in the brain increases[13]. Nonetheless, the effect of ketones themselves on appetite remains an open question.

THERMIC EFFECTS OF FOOD

Along with differences in the ways in which they affect hunger and satiety, there are also differences in how nutrients can affect the burning of calories. This quality of a nutrient is called its *thermic* effect. This is the energy the body has to expend to digest, carry, store and burn that nutrient. A component of the thermic effect is thermogenesis, which is production of heat. Thermogenesis is essential for maintaining body temperature. However, thermogenesis can also be used to regulate levels of potentially toxic fuels. When there is too much of a certain fuel, the cells can burn it off and turn it into heat instead of going through the usual processes of storing its energy in chemical form. This is one means by which the body reduces its load of fatty acid and protects itself from lipotoxicity.

Studies of the thermic effects of food have shown that the body spends more energy metabolizing protein than it spends on fat or carbohydrate. If 100 calories of protein are eaten, as many as 20 to 35 of those calories may be needed just to digest and burn that protein, as well as to carry away the waste products. This is in comparison to only 5 calories needed to metabolize 100 calories of fat or carbohydrate. Although it may be healthier to use calories to climb up a flight of stairs, the burning off of those extra calories in the processing of protein can still play a role in weight loss. It has been calculated that by simply increasing one's diet from 15–30 percent protein, with no changes in total calorie intake or exercise, an individual could lose as much as 4 lb in 1 year[14]. However, in the real world, spending a year on a diet low in carbohydrate and high in protein and fat produces weight loss very similar to that seen with diets low in fat and high in carbohydrate. Thus, advantages of the higher thermic effect of protein may not be sustainable over the long term.

THE ATKINS DIET

The best known of the low-carbohydrate diets is the Atkins diet. The Atkins diet demands an extreme reduction of carbohydrate to less than 5 percent of total calorie intake. However, the most controversial part of the Atkins diet is the virtual free-for-all intake of as much protein and fat as you desire. Fatty bacon, steaks, eggs and other cardiovascular horrors are encouraged.

The object of the Atkins Diet is to induce dependence on fat metabolism with resulting production of ketones. Atkins hypothesiszed in his book, *The Atkins Diet Revolution* published in 1971, that putting a person into "fat burning mode" increased the rate of burning of fat that had accumulated in the body. The ketones were important, because he believed that the ketones produced in the metabolism of fat, in the absence of carbohydrate, cut appetite. Moreover, because ketones in the blood are lost in the urine, he stated that urinating away the calories in ketones resulted in still further weight loss. Although an appealing idea, this loss of ketones is unlikely to contribute much to weight loss. Even strict adherence to the Atkins diet results in the loss of only a gram or so of ketones in the urine[15]. This amount of ketone would amount to a loss of less than 10 calories of fat per day, or less than 2 ounces of fat per month.

From the beginning, the Atkins Diet was seen as medical heresy. The primary concerns about the Atkins and other low-carbohydrate, high-fat, high-protein diets were stated in a 2001 position paper from The American Heart Association[16]. They noted that high protein diets generally depend upon animal sources of protein that contain high levels of saturated fat. This increase in saturated fat can raise levels of LDL. Because all protein contains amino groups, the breakdown of extra protein also increases serum levels of uric acid that can exacerbate gout and kidney stones in susceptible individuals. Another concern was what is missing from the diet when carbohydrate-containing foods are restricted. The grains, fruits and vegetables from which we obtain carbohydrate are also the primary sources of vitamins, minerals, and fiber in the diet. It was argued that increasing saturated fat and protein, while reducing the intake of carbohydrates, exposes the patient to increased risks of high cholesterol, hypertension, gout, kidney disease, osteoporosis and cancer.

Despite the dire predictions of cardiologists, one of the first scientific studies of the Atkins diet published in *The New England Journal of Medicine* in 2001[17] found the diet to be both effective and surprisingly benign. After one year, weight loss in the Atkins group was about 4 percent more than that observed in subjects on a standard, low-fat, reduced calorie diet. LDL cholesterol did not significantly differ between groups. On the other hand, as might be expected from a low-carbohydrate diet, those in the Atkins diet group had higher levels of HDL and lower levels of triglycerides. Similarly benign reports about the Atkins diet soon followed[18,19]. Admittedly, there have been no studies of the long-term effects of the Atkins diet. Nonetheless, as yet the grim consequences of the diet predicted by cardiologists have not materialized.

It is not entirely clear why the Atkins diet has advantages in bringing about weight loss. It is possible that the thermic effects of the high intake of protein and fat contribute to weight loss in the Atkins diet. However, the main reason people lose more weight on this diet may simply be that they take in fewer calories[20]. The decrease in calorie intake could be due to an appetite suppressing effect of ketones, or to the ability of large amounts of protein to reduce appetite.

Although some studies have shown weight loss advantages in the Atkins diet as much as a year after starting the diet, this advantage occurs primarily in the first few months of the diet. Moreover, while the freedom to eat as much meat, butter and fat as you please is initially enticing for many people, in the long term the Atkins diet is just as difficult to maintain as other, more standard diets. Still, what has remained the greatest barrier to acceptance of the Atkins diet by the medical community has been the unrestricted intake of saturated fat and cholesterol.

Recent publications from the Atkins Foundation reflect their recognition that many of the benefits of the Atkins Diet arises from its amelioration of the pathology of Metabolic Syndrome. Proponents of the Atkins Diet have also come to recognize that individuals who are overweight but have healthy serum triglyceride and HDL levels may not greatly benefit from restriction of carbohydrate. They have concluded that a diet should be tailored to a patient's particular metabolic profile. The Atkins Foundation has recently addressed the medical community's concerns about the high intake of saturated fat that has been associated with the Atkins diet[21]. They note that highly saturated fats and meats rich in cholesterol can easily be replaced by leaner fish, chicken, nuts, olive oil and other plant sources of fat.

OTHER LOW-CARBOHYDRATE DIETS

Many low-carbohydrate diets have been developed since the Atkins Diet burst onto the scene in 1971. Some offer the advantages of his diet without compromising what have become fairly well-established rules of healthy nutrition. The South Beach diet was developed by the Miami Beach cardiologist, Dr. Arthur Agatston. Agatston was fully aware of Gerald Reaven's work and the fact that many people begin to sufffer worsening of cholesterol and triglyceride profiles when consuming too much carbohydrate. On the other hand, as a cardiologist, he was extremely uncomfortable recommending diets that encouraged consumption of what seemed to be dangerously high amounts of saturated fat. Agatston devised a diet that reduced intake of high-glycemic index carbohydrates, such as bread, potatoes and pasta. He recommended that a significant portion of calories consumed during the day come from fruit, whole grains, vegetables, and nuts. While encouraging fairly high protein intake, he discouraged fatty meats high in saturated fat. In limiting the intake of saturated fats, he advised eating foods rich in monounsaturated fat, such as olive oil, and polyunsaturated fats such as vegetable oil. Fish high in *omega-3* fatty acids are also recommended in the South Beach diet.

Other reduced carbohydrate diets include Dr. Nicholas Perricone's Diet, the Rosedale Diet, and Dr. Barry Sears' Zone Diet. Although some rather extravagant claims are made by the authors of these diets, overall, the diets themselves are sound. They can be expected to produce weight loss and counteract the adverse effects of consuming too many high glycemic index carbohydrates.

GERALD REAVEN'S SYNDROME X DIET

In his book, *Syndrome X: Overcoming the Silent Killer That Can Give You a Heart Attack*, Dr. Gerald Reaven describes what he calls the "Syndrome X Diet".

He advises that fat intake be a relatively high 40 percent of calories, with minimal intake of saturated fats and cholesterol. He further recommends that carbohydrates make up 45 percent of the daily calorie intake, and protein contribute the remaining 15 percent of calories. Though higher in total fat, Reaven's diet is as low in saturated fat as the American Heart Association diet. The extra fat allows Reaven to reduce the amount of carbohydrate to 45 percent, whereas in the Heart Association diet, carbohydrate is what Reaven describes as "a whopping 55 to 60 percent", of total calorie intake. His is not a dramatic reduction of carbohydrate. Carbohydrate is reduced just enough to minimize the daily release of insulin that drives Metabolic Syndrome in susceptible individuals.

In Reaven's diet, it is the recommended percentages of 40 percent fat, 45 percent carbohydrate, and 15 percent protein, and not specific foods that are important. Aside from insistence on lean meat and elimination of foods high in saturated fat, such as butter and cream, there are few restrictions on the sources of the nutrients. Bread, crackers, potatoes and even sugar appear in many of his meal plans. It is the total, not the type of, carbohydrate that matters most to Reaven. He emphasizes that "there is nothing magical, exotic or complicated" about his diet, and that its strength is in being quite simple and ordinary. This way, he maintains, "you'll have no trouble sticking to it for life."

Although Reaven's Syndrome X Diet improves the manifestations of Metabolic Syndrome, it still relies on mere calorie reduction to produce weight loss. Reliance on calorie reduction alone can make dieting difficult. Although he voices skepticism about extra protein making dieting easier, there is considerable scientific evidence that the high intake of protein and fat in Atkins-type diets can make losing weight easier, at least during the first month or two. Reaven's argument against replacing carbohydrate with protein is that protein, like carbohydrate, stimulates the release of insulin. However, considering that both cause release of insulin, there is no compelling evidence of extra risk for generally healthy individuals in replacing carbohydrate with protein. Indeed, unlike the intake of carbohydrate, particularly high-glycemic index carbohydrates, protein does not produce a dramatic increase in glucose that in some people can set up a reactive hypoglycemia. The reactive hypoglycemia in turn can cause increases in hunger that lead to breaking the diet. Moreover, gram for gram, protein has fewer calories than fat, has a higher thermic effect than either fat or carbohydrate, and is more satiating. Thus, for the purpose of weight loss, replacing carbohydrate with healthy sources of protein has some clear advantages. After weight goals are achieved, Reaven's diet plan to maintain health makes more sense.

THE MEDITERRANEAN DIET

The Mediterranean Diet is a style of eating common among the people who live along the Mediterranean Coast[22]. This encompasses an area stretching from Spain and Portugal, across France, Italy, Greece, and Turkey, to the Middle East. Thus there is no single Mediterranean diet, rather there are many diets from those areas that share certain characteristics. A common feature among these diets is the liberal use of olive oil. The staples of these diets tend to be whole grains, beans, vegetables, fruits, and nuts. Moderate amounts of fish, poultry, and dairy products are consumed, but little red meat. Eggs, which are high in both cholesterol and saturated fat, are rarely consumed. Red wine with meals is another common feature of the Mediterranean diet. As I discuss in the section about resveratrol in Chapter 10 on supplements, there is some basis to suspect that red wine may offer special benefits for individuals wishing to treat or avoid Metabolic Syndrome.

In Mediterranean diets, fat makes up about 30 percent of total calories, with most of the fat coming from olive oil. Because olive oil is rich in monounsaturated fat, and the consumption of meat and dairy products high in saturated fats are minimal, the overall intake of saturated fat is low. In fact, saturated fat is typically less than 10 percent of total calories in Mediterranean diets, which meets the standards of the American Heart Association. Whereas olive oil supplies monounsaturated fatty acid, the fish, nuts and whole grains supply *omega-3* and *omega-6* fatty acids. Thus, there is ample supply of the essential fatty acids.

In some respects, the Mediterranean Diet, is similar to Reaven's Syndrome X Diet. Both include relatively modest amounts of protein. A major difference is in the reliance on calories from fat. In Reaven's diet, about 40 percent of the day's calories come from fat. In the Mediterranean diets, about 30 percent of calories come from fat, with calories from carbohydrate increased to about 55 percent. Since the carbohydrates in the Mediterranean diet are low in glycemic index and high in fiber, the extra carbohydrates do not appear to significantly exacerbate Metabolic Syndrome. In fact, there is evidence that eating a Mediterranean style diet can improve serum glucose levels and reduce the insulin resistance of Metabolic Syndrome[23].

The Meditarranean style of eating offers a variety of health benefits. The Lyon Diet[24], which is the best studied of the various Mediterranean diets, has been found to reduce cardiovascular risk. The Lyon Diet was the subject of a large, 4 year study of the effects of diet on survival rates in individuals who had already suffered a heart attack. After 4 years, patients eating the Lyon Mediterranean Diet had a 55 percent reduction in risk of death, and a nearly 70 percent decrease in rates of recurrence of heart attacks. Mediterranean style diets have been reported to reduce the risk of developing Alzheimer's Disease[25]. The Meditrranean diet can also reverse some of the sexual dysfunction in men and women attributed to the vascular disturbances caused by Metabolic Syndrome. These effects are thought to be benefits of the monounsaturated fatty acid in the olive oil. Weight loss on the Mediterranean diet is still a matter of struggling to cut back on calories. The reduction of high glycemic index carbohydrates may help some individuals avoid the hunger that arises from postprandial hypoglycemia. There is nothing in the diet that enhances satiety or thermic effects of food. Still, the Mediterranean style of eating appears to be an excellent maintenance diet, with many beneficial effects.

THE BOTTOM LINE ON DIET

A good weight loss diet should improve overall health, and not sacrifice health simply to lose a few pounds. If Metabolic Syndrome is a concern, the diet should enhance insulin sensitivity, and help reverse the high LDL, low HDL, hypertriglyceridemia, hypertension, and inflammation that are the consequences of the syndrome. The diet should be pleasant, safe and convenient enough to continue as long as necessary, while flexible enough to allow adjustments to fit the changing needs of the dieter.

The only diets that make weight loss a little easier, albeit for only the first few months, are the low-carbohydrate, high-fat and high-protein diets, such as the Atkins Diet. The Atkins Diet, which is the most extreme form of this type of diet, is not nearly as malignant as had been expected. Nonetheless, this should not be seen as a ringing endorsement. There is much that can be improved in the Atkins type of diet to minimize legitimate concerns about the high intake of saturated fat and the loss of important nutrients, such as fiber, vitamins, and minerals, that results from reduced consumption of plant sources of carbohydrates. Even Gerald Reaven himself admitted that, if the enormous load of saturated fat in the Atkins diet could be reduced, the diet could be very helpful for those with Metabolic Syndrome.

There are many easy ways to reduce the saturated fat in an Atkins type diet. Instead of steak, salmon or other fish high in *omega-3* fatty acids can be eaten. Beef can also be replaced with buffalo meat that is significantly leaner and lower in saturated fat. Although buffalo was once an exotic meat, it is readily available in most cities. Instead of regular eggs, egg substitutes or *omega-3* fatty acid enriched eggs can be consumed at breakfast. Although bacon and eggs is standard American fare, kippers or lox can be tried instead. Butter can be replaced by one of the modern soft margarines that are free of *trans* fats and low in saturated fats. Adding dairy products to the diet can be a good source of protein and may possibly be helpful for weight loss[26]. Low fat dairy products, such as 2 percent milk and cottage cheese, are rich in protein while having acceptable levels of saturated fat. There are even several brands of reduced carbohydrate milk available.

Because low carbohydrate diets tend to be deficient in vitamins, minerals and fiber, supplements are worthwhile, if not essential. In the following chapter, I will discuss a number of supplements that can be added to the diet for specific reasons. Nonetheless, it is safe to say that anyone on a low-carbohydrate diet should at minimum supplement their nutrient intake with a good multivitamin; antioxidant vitamins C and E; calcium; magnesium; chromium; fiber; and lots of good, pure water.

Because the advantages of a high-protein, high-fat diet are fairly short-lived, adjustments should be made to return healthy forms of carbohydrate back into the diet after the first few months. At that time, it might be worthwhile to switch to one of the appealing Mediterranean style diets.

REFERENCES

1. Howard B.V., et al., Low-fat dietary pattern and risk of cardiovascular disease: the women's health initiative randomized controlled dietary modification Trial. *JAMA* 2006; 295:655–666.
2. Conn J.W. and Newburgh, L.H., The glycemic response to isoglucogenic quantities of protein nd and carbohydrate. *J. Clin. Invest.* 1936; 15:667–671.

3. Nuttall, F.Q. and Gannon M.C., Plasma glucose and insulin response to macronutrients in nondiabetics and NIDDM subjects. *Diabetes Care* 1991; 14:824–838.

4. Messner, M. and Brissot, P., Traditional management of liver disorders. *Drugs* 1990; 40(Suppl. 3):45–57.

5. Lentine, K. and Wrone, E.M., New insights into protein intake and progression of renal disease. *Curr. Opin. Nephrol. Hypertens.* 2004; 13:333–336.

6. Choi, H.K. et al., Purine-rich foods, dairy and protein intake, and the risk of gout in men. *N. Engl. J. Med.* 2004; 350:1093–1103.

7. Halton, T. L., and Hu, FB., The effects of high protein diets on thermogenesis, satiety and weight loss: a critical review. *J. Am. Col. Nut.* 2004; 23:373–385.

8. Long, S.J. et al., Effect of habitual dietary protein intake on appetite and satiety. *Appetite* 2000; 35:79–88.

9. Ludwig, D.S. et al., High glycemic index foods, overeating and obesity. *Pediatrics* 1999; 103:261–266.

10. Bowen, J., Noakes, M., and Clifton, P.M., Appetite regulatory hormone responses to various dietary proteins differ by BMI status despite similar reductions in ad libitum energy intake. *J. Clin. Endocrinol. Metab.* May 30, 2006 (Epublished ahead of print).

11. Rosen, J.C. et al., Mood and appetite during minimal-carbohydrate and carbohydrate-supplemented hypocaloric diets. *Am. J. Clin. Nutr.* 1985; 42:371–379.

12. Veech, R.L., The therapeutic implications of ketone bodies: the effects of ketone bodies in pathological conditions: ketosis, ketogenic diet, redox states, insulin resistance, and mitochondrial metabolism. *Prostaglandins Leukot. Essent. Fatty Acids* 2004; 70:309–319.

13. Yudkoff, M. et al., Brain amino acid metabolism and ketosis. *J. Neurosci. Res.* 2001; 66:272–281.

14. Eisenstein, J. et al., High proteinweight loss diets: are they safe and do they work? A review of the experimental and epidemiological data. *Nutr. Rev.* 2002; 60:189–200.

15. Westman, E.C. et al., Effect of 6-month adherence to a very low carbohydrate diet program. *Am. J. Med.* 2002; 113:30–36.

16. Sachiko, T. et al., Dietary protein and weight reduction. *Circulation* 2001; 104:1869.

17. Foster, G.D. et al., A randomized trial of a low-carbohydrate diet for obesity. *N. Engl. J. Med.* 2003; 348:2082–2090.

18. Stern, L. et al., The effects of low-carbohydrate versus conventional weight loss diets in severely obese adults: one-year follow-up of a randomized trial. *Ann. Intern. Med.* 2004; 140:778–785.

19. Yancy, W.S. et al., A low-carbohydrate, ketogenic diet versus a low-fat diet to treat obesity and hyperlipidemia. *Ann. Intern. Med.* 2004; 140:769–777.

20. Boden, G. et al., Effect of a low-carbohydrate diet on appetite, blood glucose levels, and insulin resistance in obese patients with type 2 diabetes. *Ann. Intern. Med.* March 15, 2005; 142:403–411.

21. Bloch, A.S., Low carbohydrate diets, Pro: Time to rethink our current strategies. *Nutr. Clin. Prac.* 2005; 20:3–12.

22. Cloutier, M. and Adamson, E., *The Mediterranean Diet.* Avon Books, New York, 2004.

23. Ryan, M. et al., Diabetes and the Mediterranean diet: a beneficial effect of oleic acid on insulin sensitivity, adipocyte glucose transport and endothelium-dependent vasoreactivity. *QJM* 2000; 93:85–91.

24. de Lorgeril, M. et al., Mediterranean diet, traditional risk factorrs and the rate of cardiovascular complications after myocardial infarction: Final report of the Lyon Diet Heart Study. *Circulation* 1999; 99:779–785.

25. Scarmeas, N. et al., Mediterranean diet and risk for Alzheimer's Disease. *Ann. Neurol.* 2006; 59:912–921.

26. Zemel, M.B., The role of dairy foods in weight management. *J. Am. Coll. Nutr.* 2005; 24(6 Suppl.):537S–46S.

10
NUTRITIONAL SUPPLEMENTS AND METABOLIC SYNDROME

ACETYL-CARNITINE

Acetyl-carnitine is a substance natural to the body. It is readily formed in cells by the enzymatic addition of an acetyl group to carnitine. Carnitine, also natural to the body, is a modified version of the amino acid lysine. Enzymes can readily convert carnitine to acetyl-carnitine and back, according to the metabolic needs of the cell. Thus, inside the cell, acetyl-carnitine and carnitine are essentially interchangeable. The major difference between acetyl-carnitine and carnitine is that acetyl-carnitine is more easily absorbed from the gut, and more readily crosses the blood-brain barrier. As a supplement, acetyl-carnitine has certain advantages.

Carnitine serves an important role in the burning of fat for energy, and ferrying fatty acids across the walls of the mitochondria, where they are oxidized and turned into energy. On its return trip from inside the mitochondria, it brings back acetyl groups and other small fragments produced in fatty acid oxidation. Carnitine's role in fat metabolism leads it to affect, in some degree, all the other energy metabolism in the cell, including the burning of carbohydrate. In this way it can be linked to glucose metabolism, insulin and Metabolic Syndrome. Some authors have suggested that its important role in the burning of fat also makes it useful in diets for weight loss and overall health maintenance.

We get a constant supply of carnitine in our diet, especially if we eat meat. (In fact, the name carnitine is derived from the Latin word for meat.) Although most diets supply more than enough carnitine, our bodies have the capacity to make our own supply through enzymatic action on the amino acid lysine. By that process even strict vegetarians are thought to be able to supply themselves with adequate amounts of carnitine[1]. There are some hereditary illnesses in

Metabolic Syndrome and Psychiatric Illness: Interactions, Pathophysiology, Assessment & Treatment
Copyright © 2008 by Academic Press. All rights of reproduction in any form reserved.

which the production or absorption of carnitine is deficient; however, those illnesses are rare.

Because most of us eat it, and all of us make it, it is not entirely clear if it is necessary or even worthwhile to supplement the diets of healthy people with carnitine or acetyl-carnitine. There may be circumstances in which supplementation could be beneficial. For example, it has been found that carnitine levels decrease significantly as we age[2]. There is compelling evidence that 1.5–3 g a day of acetyl-carnitine can improve cognitive function of individuals in the milder, early stages of Alzheimer's Disease[3]. It is not clear if acetyl-carnitine improves cognition by giving mitochondrial activity in the brain a boost, or by making more acetyl groups available to be used in the synthesis of the neurotransmitter acetylcholine in the brain. Acetylcholine levels are decreased in many forms of dementia, and most of the drugs used for treatment of Alzheimer's Disease increase the availability of acetylcholine by preventing its enzymatic destruction in the brain.

There have recently been a number of reports that acetyl-carntine can be helpful in the treatment of depression in elderly patients[4]. In most cases, the elderly patients in the studies had both depression and some type of dementia. However, in one study, the patients were depressed but did not have dementia. Thus, the beneficial effects of acetyl-carnitine are not due entirely to improvement of compromised cognitive function. Acetyl-carnitine and the closely related substance propionyl-carnitine, often used in Europe but not in the United States, may also be useful in the treatment of chronic fatigue syndrome[5].

Enzymes that metabolize fat in the cell are always in chemical communication with those enzymes metabolizing glucose. Such communication may explain how intravenously administered carnitine is able to improve serum glucose levels and utilization in patients with diabetes type II[6]. Another relevant finding is that acetyl-carnitine can ease the pain and abnormal function in patients with diabetic peripheral neuropathy[7]. Oxidative stress and the four primary pathways of damage that it creates, so well described by Dr. Brownlee, are primarily responsible for neuropathic damage in diabetes[8]. Recent evidence suggests that acetyl-carnitine may offer some protection from oxidative damage, and this protection may even extend to the brain[9]. As you may recall, oxidative stress is thought to play a role in the progression of Alzheimer's Disease in humans.

Animal studies have shown that giving acetyl-carnitine to very old rats lowers the concentrations of cholesterol and triglycerides in their blood[10]. No such studies have been performed in healthy, elderly human beings. It is worth noting that giving 1 g a day of carnitine to patients on dialysis due to severe kidney disease, has been reported to reduce triglycerides and LDL cholesterol while increasing HDL cholesterol[11]. However, dialysis itself greatly reduces serum levels of carnitine. That may be more an indication of the adverse effects of carnitine depletion in seriously ill patients than a basis to begin carnitine supplementation in people with mild or even moderate concerns about their health.

There is evidence from both human and animal studies that carnitine, or its more readily absorbed form, acetyl-carnitine, can to some degree improve glucose tolerance, insulin sensitivity, serum lipids, and cognitive function. It may also offer some protection from oxidative damage. Depression and fatigue can be improved. Perhaps the best feature of acetyl-carnitine is that it is a natural substance with very little likelihood of producing adverse effects.

ALPHA-LIPOIC ACID

Alpha-lipoic acid (ALA) is a natural substance in the human body. It is obtained to some degree in our diet, coming primarily from organ meats. It is also present in plant sources, but at much lower levels. We also synthesize our own ALA in the liver and other tissues. However, some researchers suspect that synthesis of ALA is barely adequate, and that supplementation is useful.

Like many substances of organic origin, ALA can exist in right- and left-handed, mirror imaged forms called stereo-isomers. It is the right-handed form (R-ALA) that is natural to the body and biologically active. L-ALA is found in synthetic ALA, which is a 50/50 mixture of the two stereo-isomers. L-ALA is not only inactive, but is also suspected of hindering the activity of natural R-ALA.

R-ALA serves a number of functions in the human body. Primarily it is a cofactor in the activities of two important enzymes in the mitochondria. These enzymes, pyruvate dehydrogenase and alpha-ketoglutarate dehydroegnase play fundamental roles in glucose metabolism. Its ability to participate in oxidation-reduction reactions makes it well-suited to serve that purpose in the mitochondria. However, this property also makes it an extremely potent and important antioxidant molecule. As an antioxidant, R-ALA has been found to reduce a variety of sources of oxidative stress in the body. It is also thought to reactivate vitamins E and C after they have been used as antioxidants. Whereas vitamin E is soluble in fat and vitamin C is soluble in water, R-ALA is soluble in both fat and water. Thus, it has full access to all tissues of the body, including the brain, where it can act as an antioxidant.

One therapeutic use of R-ALA has been the treatment of painful diabetic neuropathy. It has been used in Europe for this purpose for over 30 years. It is not entirely clear what led to the suspicion that it might also be useful in treating the more primary metabolic components of diabetes. Perhaps it was a serendipitous observation of improvements in diabetics being treated for neuropathy. In any case, it has been found that oral dosing of 600 mg of R-ALA per day significantly improves insulin sensitivity in diabetic patients[12].

It is not entirely clear how R-ALA acts to enhance the activity of insulin. It may independently stimulate some of the chemical pathways that insulin itself uses to produce its effects upon cell activity[13]. R-ALA increases uptake of glucose into cells partly through enhancing the activity of pyruvate dehydrogenase and, in turn, the ability of cells to use glucose as fuel. Moreover, like insulin, it inhibits gluconeogenesis in the liver and stimulates uptake of glucose by GLUT4 receptors in cell membranes[14]. Thus, R-ALA may act in a variety of ways to enhance insulin's effects and reduce the insulin resistance of Metabolic Syndrome.

Administration of ALA prevents the progression of diabetes in a strain of laboratory rats that spontaneously develop this illness[15]. This effect of

ALA may be due to its prevention of accumulation of triglycerides in the muscle cells and pancreases of these animals. Triglyceride accumulation in muscle cells leads to insulin resistance, whereas accumulation in the pancreas leads to oxidative damage that prevents the beta cells from keeping up with increases in demand for insulin. R-ALA also reduces serum triglyceride levels[16] in laboratory rats, as would be expected of a substance that mimics or enhances the effects of insulin.

Both the lipotoxicity and glucotoxicity that injure the pancreatic beta cells are due in part to damage from oxidation. The potent antioxidant effects of R-ALA may spare the pancreas and other organs damage from the out-of-control oxidation seen in Metabolic Syndrome. However R-ALA may have even more far-reaching ability to spare the body from damage. R-ALA also reduces the ability of the immune cell messenger and adipocytokine TNF-alpha to induce NFk-B[17]. NFk-B is a chemical switch that initiates some important inflammatory processes. It sets into motion the production of a variety of substances, including many of the cytokines and adipocytokines that promote Metabolic Syndrome. R-ALA may help break the vicious cycle of oxidative damage, inflammation, weight gain, insulin resistance, and further weight gain that drives the Metabolic Syndrome to spiral out of control.

It remains to be determined if R-ALA can prevent or reverse diabetes in humans. Nonetheless, the insulin-enhancing, antioxidant, and anti-inflammatory effects of R-ALA would appear to make it a very useful substance. Moreover, there appears to be little if any toxicity of the substance. Because R-ALA competes with the B vitamins biotin and pantothenic acid for uptake into cells, it has been suggested that supplementation with R-ALA should also include supplementation with the B-complex of vitamins.

CARNOSINE

Carnosine, like carnitine, comes primarily from meat. However, while their names are similar, their actions in the body are quite different. Carnosine is useful in helping prevent damage caused by too much sugar in the body[18]. Sugars, such as glucose or fructose, can bind with proteins. This binding not only damages the protein, but the glycosylated protein can stimulate secondary inflammatory processes. Glycosylation of protein is one of the four major destructive pathways stimulated by oxidative stress in the body. The binding of sugars to protein can also lead to proteins sticking to each other, a process known as cross-linking. The destructive effects of protein glycation and cross-linking play major roles in the ageing process. For example, cataracts that tend to form in the lenses of the eyes as we age are thought to be due to cross-linking of proteins secondary to glycation. This is the reason that cataracts are so common among sufferers of diabetes.

Carnosine is formed from the binding together of the amino acids alanine and histidine. This pair of amino acids presents a binding site for glucose and other sugars that is very similar to sites where sugars bind on complete proteins. In individuals with hyperglycemia, either from Metabolic Syndrome or fullblown diabetes, glycosylation of proteins accelerates. In people with diabetes, carnosine levels are lower than in people without diabetes[19], perhaps because it gets used up in its role of binding to excess glucose in the blood. Strict vegetarians, who do not consume much carnosine in their diet, tend to have higher levels of glycosylated proteins in their bodies than do people who eat meat[20].

Although it has long been known that carnosine helps reduce damaging protein glycation in the body, scientists have reported other potentially beneficial effects of carnosine as well. For example, carnosine has excellent antioxidant properties. It can greatly reduce damage from the highly reactive oxygen free radicals that are produced from the out-of-control oxidation of lipids and sugar that can occur in Metabolic Syndrome and diabetes[21]. Moreover, this antioxidant effect occurs at concentrations of carnosine known to exist in the human body. In animal studies, administration of high doses of carnosine reduced oxidative damage in the brain and blood during periods of high stress[22]. In a study using rats made salt-sensitive with stress hormones, intravenous administration of carnosine decreased their blood pressure and reduced sympathetic nervous system activity in their kidneys[23].

In the human body, concentrations of carnosine are highest in muscle tissue. However, brain tissue also has fairly high concentrations[24]. The brain may be in particular need of the antiglycosylation and antioxidant effects that carnosine provides. Glycosylation, oxidative stress, inflammation secondary to the oxidative stress, and protein cross-linking from methylglyoxal, a product of uncontrolled glucose oxidation, are all thought to contribute to the progression of Alzheimer's Disease. There is evidence from animal studies that carnosine can reduce the toxic effects of beta-amyloid protein[25]. Beta-amyloid is the abnormal protein that builds up in the brains of people with Alzheimer's Disease and damages the surrounding brain tissue. For these reasons, increasing levels of carnosine in the brain may have a protective effect from Alzheimer's Disease.

Unfortunately, there are questions about the efficacy of oral supplementation with carnosine. Most carnosine absorbed from the gut is destroyed in the bloodstream by enzymes called carnosinases. These enzymes readily split carnosine back into its two constituent amino acids, alanine and histidine. This fact has led to trials of intranasal administration of carnosine. This might be a very clever way to get it into areas of the brain, such as the olfactory lobes, that are ravaged by glycosylation and oxidative damage in sufferers of Alzheimer's Disease[26]. Nonetheless, there is evidence from human studies that supplementing with carnosine can increase carnosine levels in muscle tissue[27]. Supplying people with the amino acids the body uses to make carnosine also increases carnosine levels in muscle. Therefore, even if enzymes do break carnosine apart in the blood, the pieces can be put back together in the cells.

Unless one is a strict vegetarian, carnosine supplementation is likely unnecessary. Carnosine levels in the blood increase after a person has a meal of beef[28]. Whereas a 1000 mg of carnosine a day has been recommended as a supplement, there is about 1500 mg of carnosine in a pound of beef, and close to 2000 mg in similar amounts of pork or chicken[29]. Most fish, such as salmon, are low in carnosine, but high in a substance called anserine. Anserine is also found in the human body, and has actions in cells quite similar to those of carnosine[30].

CHOCOLATE

Getting healthy does not entail deprivation. This is clear from the fact that chocolate is very good for you. Chocolate was cultivated by the Olmec Indians of Central America as far back as 1500 BC. It was difficult to obtain and jealously guarded. It remained a drink of the Indian aristocracy. After it was brought back to Europe by the Spaniards, it became a drink of the European elite. In 1753, the great Swedish taxonomist, Linnaeus, quite rightfully gave the cocoa plant the scientific name *theobroma*, which means "food of the gods". It was not until the 1800s that chocolate became accessible for common folk. Better late than never.

Chocolate is chockfull of substances known as flavonoids. Flavonoids are multiringed molecules found in a variety of plants. They give many fruits and vegetables their dark, rich colors. It turns out that chocolate has more helpful flavonoids than most foods, including teas or the highly touted red wine.

Flavonoids in chocolate are powerful antioxidants. Habitual intake of chocolate can reduce cardiac risk by nearly 20 percent through its ability to decrease LDL oxidation, increase HDL, reduce inflammation, decrease blood pressure, and dampen platelet activity[31]. Chocolate also reduces insulin resistance[32].

Many of the benefits of chocolate are from its antioxidant effects. Oxidation not only causes direct damage to tissues, but it also triggers inflammatory processes that cause secondary damage. However, there appear to be benefits from chocolate that are not simply the results of antioxidant effects. For example, chocolate enhances the effects of nitric oxide by stimulating the activity of nitric oxide synthase[33]. Nitric oxide is a potent vasodilator. Thus, increases in nitric oxide probably mediate the antihypertensive effects of chocolate. One suspects that chocolate might also improve sexual function in sufferers of Metabolic Syndrome by increasing blood flow in penile and clitoral vasculature. It is also worth noting that some of the sexual dysfunction from antidepressants, particularly the SSRIs, may be due to inhibition of nitric oxide synthase[34]. To the best of my knowledge, there have been no studies of the ability of chocolate to reverse the sexual side effects of antidepressants.

Some benefits of chocolate flavonoids may come from their effects on eicosanoids. Consumption of chocolate tends to inhibit synthesis of leukotrienes. Leukotrienes act to constrict blood vessels, promote inflammation, and activate platelets. At the same time, chocolate enhances synthesis of prostacyclins, which tend to enhance microvascular blood flow, prevent inflammation, and calm platelet activity[35].

Chocolate contains significant amounts of phenylethylamine[36]. This substance, which is the decarboxylation product of the amino acid phenylalanine, has long been suspected of having antidepressant effects[37]. In recent years, it has been discovered that phenylethylamine is likely to be the endogenous ligand at the Trace Amine-Associated Receptor 1 (TAAR1) in brain tissue[38].

It binds with high affinity to that receptor. Interestingly, TAAR1 is also one of the sites of action of methamphetamine. This may explain some of the "addictive" quality of chocolate. Chocolate also contains theobromine and small amounts of the related substance caffeine. Theobromine is not nearly as good a stimulant as caffeine, thus the stimulant effects of chocolate are probably not significant. A fascinating finding is that chocolate contains small amounts of anandamide[39]. Anandamide is an endocannabinoid, which is a term coined from the phrase "endogenous cannabinoid". Endocannabinoids are the natural substances in the brain that marijuana mimics. The amount of anandamide in chocolate is quite small. However, other substances in chocolate similar in structure to anandamide may act on the enzymes systems in the brain to increase the levels of endogenous cannabinoids in the brain.

I must add that the brain is well-insulated from active amines in the blood, and there are serious questions as to whether the phenylethylamine and other psychoactive compounds in dietary chocolate have any important impact on brain activity. A fascinating study performed in 1994 casts doubt on this possibility[40]. White chocolate without the psychoactive compounds of dark chocolate gave relief from "chocolate craving", whereas addition of the psychoactive compounds in capsule form did not add to that effect.

The dark side of chocolate is the amount of sugar and saturated fat that it usually contains. The first of these problems is easily solved. There are many delicious, sugar free dark chocolates on the market now. The second problem might not be as bad as it first may seem. The cocoa butter mixed into chocolate to form bars contains about 30 percent oleic acid, which is the healthy, monounsaturated fatty acid in olive oil. Cocoa butter is rich in saturated fatty acids. It contains about 25 percent palmitate and 33 percent stearate. Admittedly, aside from being a natural product of our metabolism, palmitate has few redeeming features. On the other hand, as I have mentioned in the chapter on fats, stearic acid has probably gotten a bad rap. Stearic acid tends not to increase LDL or decrease HDL. The daily feeding of as much as 10 oz of milk chocolate bars does not adversely affect serum cholesterol profiles[41].

CHROMIUM

In 1959 Dr. Walter Mertz was the first to suggest that chromium is an essential trace mineral, and that it plays a role in enhancing the activity of insulin[42]. Dr. Mertz, who died in 2002, was the director of the US Department of Agriculture, Human Nutrition Research Center from 1972 until his retirement in 1993. He was considered a world authority on the role of trace minerals in human nutrition.

Since Dr. Mertz' first reports, researchers have accumulated a large and compelling body of literature indicating that chromium enhances the activity of insulin. However, many questions remain about the use of chromium. There is still some controversy about whether chromium is truly an essential mineral, or whether it simply acts as a drug to increase the body's sensitivity to insulin. There are also continuing concerns about possible toxicity of chromium, and over what form and dosage of chromium is safest and most effective for combating insulin resistance.

Although there are scientists still voicing doubts, there appears to be very strong evidence that chromium is essential for human health[43]. It has been found, for example, that patients with diets lacking in chromium begin to exhibit the hyperglycemia and increases in insulin levels that define insulin resistance. Moreover, while those problems are not reversed with extra insulin, they do respond to addition of chromium back into the diet. Questions remain about the exact mechanism by which chromium enhances insulin activity. A special protein called chromodulin exists in the human body that binds chromium and carries it to the insulin receptor in the cell membrane[44]. Specific measurements have been made of increases in insulin receptor activity in direct relation to increases in the amount of chromium being carried by the chromodulin.

Over the last 25 years or so, there have been a variety of studies showing that supplementation with chromium, usually in the form of chromium picolinate in doses of 200–1000 μg a day, lowers insulin resistance and improves glucose tolerance[45]. Chromium has also been reported to lower serum triglycerides[46] and increase HDL cholesterol[47] in patients with diabetes type II. Not all studies have shown chromium to affect insulin resistance and other aspects of Metabolic Syndrome[48]. However, it has been argued that negative results only indicate how important it is to use an appropriate dose of chromium in such studies.

Even proponents of the use of chromium to improve insulin sensitivity admit that the amount of chromium necessary to produce significant improvements in insulin activity makes it seem unlikely that the benefits are simply the results of correcting deficiencies. There have been statements, some even coming from the US Department of Agriculture, suggesting that chromium deficiency is common in the United States. There have been estimates that as many as 90 percent of Americans receive less than the recommended daily allowance of chromium. However, most authorities argue that chromium deficiency is rare. The US Department of Agriculture has even lowered the recommended daily requirement of chromium to be more in line with current scientific thought on the matter.

In some cases, circumstances may create functional chromium deficiencies. For example, too much iron in the body may prevent transport and effective use of chromium. The elderly may not utilize chromium efficiently. The most important data are from animal studies showing that under conditions of hyperglycemia and increased demand for insulin, such as occur in Metabolic Syndrome, there is an increase in loss of chromium in the urine[49].

There is a distinct possibility that the high doses of chromium needed to help reduce insulin resistance reflect the mineral acting more as a drug than as an essential nutrient. There is nothing about a pharmacological rather than a physiological mechanism of the action of chromium that would make it unacceptable for use in helping to reduce insulin resistance. However, if it is acting as a drug, then we must more thoroughly question the risks versus the benefits that treatment with chromium would entail.

Chromium, like many metals, can exist in several states, called valencies. The hexavalent form of chromium is extremely toxic, but this is not the type of chromium found in chromium supplements. Perhaps the most worrisome data suggesting that chromium supplementation could be dangerous are results of an experiment showing that exposure to chromium picolinate causes sterility and lethal mutations in fruit flies[50]. Although quite distant from humans on the evolutionary tree, the fruit fly is often used as a test animal to screen substances for potential to damage DNA. However, in that study, chromium picolinate, and not chromium chloride, produced damage in fruit flies. Thus, it could not have been the chromium that caused the damage. The authors suggested that the means used by the body to transfer the chromium from the picolinate to chromodulin may have produced dangerous free radicals in the fruit flies that in turn damaged their DNA. In a study of the effects of chromium on hamster ovary cells, chromium in the form of picolinate again caused damage whereas chromium in the form of chloride or nicotinate did not[51]. However, the fairly common dose of 400 μg of chromium picolinate daily for 8 weeks was found to have no damaging effects on the DNA of the women taking it[52].

If a patient is anxious about the use of chromium picolinate, it might be prudent to use chromium nicotinate at a slightly higher dose. The equivalent to 400 mg of chromium picolinate would be about 800–1000 mg a day of chromium nicotinate. Chromium-rich yeast is another reasonable alternative.

COENZYME Q10

Coenzyme Q10 is another substance natural to the body and essential for life. It is part of the chain of electron transport molecules in the mitchondria, and it is necessary for the production of energy. Another name for Coenzyme Q10 is ubiquinone, which reflects the fact that it is found in every cell of the body. Its role in accepting and transferring electrons in the mitochondria also makes it perfectly suited to act as an antioxidant, particularly in the mitochondrial environment, where many high energy biochemical reactions occur.

Because Coenzyme Q10 is synthesized in the body, it is not considered an essential nutrient. The synthesis of Coenzyme Q10 and cholesterol share many similar steps and rely on some of the same enzymes. Thus, the statin drugs used to lower high cholesterol may also lower levels of Coenzyme Q10. Although this fact was reported in the prestigious *Proceedings of the National Academy of Sciences*[53], it is rare that a doctor supplements with coenzyme Q10 when prescribing a statin.

In many respects, Coenzyme Q10 is like alpha-lipoic acid described previously. However, while there is strong evidence that alpha-lipoic acid improves insulin resistance, glucose tolerance, and inflammation, the reports of coenzyme Q10's effects on these aspects of Metabolic Syndrome have been mixed. Some studies have revealed no benefits from using Co-enzyme Q10 in Metabolic Syndrome. On the other hand, a relatively small dose of 60 mg Coenzyme Q10 twice a day for 8 weeks was reported to reduce levels of fasting insulin, glucose, triglycerides, and evidence of oxidative damage to fats[54]. Moreover, the ratio of coenzyme Q10 to ubiquinol, which is essentially used-up coenzyme Q10, has been reported to decrease in Metabolic Syndrome[55].

That finding has been taken to indicate that the body suffers increasing oxidative stress in the syndrome.

From existing data, there is no strong basis to recommend coenzyme Q10 as part of a treatment for Metabolic Syndrome. However, it is a safe and natural substance with many potential benefits. One such benefit may include its ability to slow down the progression of certain degenerative neurological diseases that are exacerbated by Metabolic Syndrome[56]. These illnesses are often associated with disturbance of mitochondrial function. Thus, in the final analysis, supplementing with 100–200 mg a day of Coenzyme Q10 is quite reasonable. Certainly, if anyone is prescribed a statin drug to reduce cholesterol, it would be wise to add coenzyme Q10.

CURCUMIN

Curcumin, a component of the spice, turmeric, has recently been touted in the lay literature to be a substance of great promise in the treatment of inflammatory illnesses, diabetes, and Metabolic Syndrome. Certainly, a natural substance acting in a multitude of beneficial ways without obvious ill effects is a very appealing possibility. Unfortunately, there are no comprehensive clinical studies on the effects of curcumin on any human disease.

Evidence in the literature consists primarily of animal studies, with a few studies using human cells. The intial curcumin data is encouraging. In diabetic rats, curcumin reversed much of the liver dysfunction that led to hyperglycemia[57]. Curcumin also lowers cholesterol and triglycerides in diabetic rats[58]. Of substantial interest is the report, again from rat studies, that curcumin stimulates PPAR-gamma, the activity of which is thought to greatly benefit Metabolic Syndrome and diabetes II[59].

Curcumin has anticancer and antioxidant effects[60], as well as anti-inflammatory properties due to its ability to block the activities of NFk-B, cyclooxygenase and nitric oxide synthase[61]. Cyclooxygenase inhibitors have usefulness in the treatment of arthritic conditions. Most prescribed antidepressants have the ability to block nitric oxide synthase activity in the brain, and some authors suspect that this effect is at least partially responsible for their ability to improve mood.

Some fascinating new evidence shows that curcumin may also have usefulness in the treatment of Alzheimer's Disease. In the test tube, curcumin appears to break up the aggregations of amyloid fibers. Most impressively, when injected into aged rats with Alzheimer's-like plaques of amyloid, those plaques begin to reduce in size[62]. In the UCLA laboratory of Dr. Jeffrey Cummings, one of the world's authorities on dementia, human trials using curcumin for treatment of Alzheimer's have already begun[63]. In view of a relation between Metabolic Syndrome and Alzheimer's Disease, it would seem that this apparently benign substance would be worth trying in any individual with Metabolic Syndrome and risk factors for Alzheimer's Disease as well.

Safety concerns have been addressed at this point in time, and existing evidence indicates that curcumin is fairly safe in humans[64]. There have been reports that high doses can cause stomach ulcers in rats[65]. Thus, it would seem prudent for people with a history of gastric ulcers to avoid circumin. In most cases, however, even enormous doses of circumin fed to laboratory rats have been found to do little more than turn their fur yellow. Perhaps the most difficult limiting factor in the use of curcumin is that it is very poorly absorbed. Perhaps a way will be found to provide therapeutic blood levels of the substance.

DHEA

DHEA is a fascinating substance that is available over-the-counter in almost every neighborhood drug store. If it were just now made available, it would almost certainly be by prescription only. DHEA stands for dehydroepiandrosterone. It is a natural substance in the body produced by the adrenal glands, and it is a member of the steroid family. It is an unusual substance in that the body can convert it into a variety of other steroids, including both testosterone and estrogen. Blood levels of the hormone peak around the age of 20, and they begin to drop precipitously after the age of 25. Some rather sobering studies have found that low levels of DHEA correlate with an increased risk of dying in the subsequent 5–10 years[66]. It should be noted, however, that there is no compelling evidence that restoring DHEA levels in such individuals serves in any way to extend life. Many benefits in mood and overall health have been attributed to treatment with DHEA, and there is quite strong evidence that DHEA is useful.

DHEA has been found to reverse some features of Metabolic Syndrome. Modest doses as low as 25 mg a day enhance insulin sensitivity and reduce inflammation, as measured by reduction in serum levels of the pro-inflammatory cytokine plasminogen activator inhibitor. Similar effects are observed in young[67] and elderly[68] patients.

Visceral fat is also significantly reduced with DHEA treatment. This may reflect a physiological antagonism between DHEA and cortisol. In fact, DHEA antagonizes a number of effects of cortisol[69]. Reduction of visceral fat and improvement of insulin sensitivity by DHEA might also be the result of stimulation of PPAR[70]. Many of the newest and most effective medications for diabetes type II act by stimulating PPAR.

A potentially important effect of DHEA is the ability to stimulate neurogenesis[71]. It used to be thought that neurons in the adult brain are not replaced after they die. However, in recent years it has been shown that the birth of new neurons, that is neurogenesis, does occur in the brains of adult mammals. In rats, corticosterone, the rodent equivalent of the stress hormone cortisol in humans, causes suppression of neurogenesis. However, DHEA blocks that effect of corticosterone in rats. Many neuroscientists suspect that some of the ability of antidepressants to improve mood is due to their ability to enhance neurogenesis. This could be part of the mechanism by which DHEA produces antidepressant effects in men and women with Major Depressive illness[72]. In women, though not so much in men, DHEA treatment also produces increases in serum testosterone. This increase results in improvement in sex drive, and may further contribute to a sense of well-being[73].

Because of its ability to antagonize cortisol, increase insulin sensitivity, and reduce inflammation, there is reason to suspect that DHEA might also be useful in the treatment of Alzheimer's Disease. Indeed, there is even evidence that a metabolite of DHEA, dehydroepiandrosterone sulfate, enhances the activity of the neurotransmitter acetylcholine in the brain[74]. Most of the FDA approved medications for Alzheimer's Disease, including Aricept, are thought to act primarily by increasing the availability of acetylcholine in the brain. Unfortunately, controlled studies on the use of DHEA for treatment of Alzheimer's have been disappointing. *The Cochrane Review*, a British organization that reviews and evaluates medical treatments, did not feel that there was sufficient proof to use DHEA in the treatment of Alzheimer's Disease. They did note that prevention of Alzheimer's by using DHEA supplementation before serious progression of Alzheimer's might still be worth investigating.

In some individuals, antidepressant treatment can cause a switch from depression to the irritability, agitation, and restlessness of mania. In fact there have been several reports of DHEA-induced mania[75]. Anyone with a personal or family history of Bipolar Affective Disorder should be extremely cautious in using DHEA. It would be prudent to use it only under a doctor's supervision. Another word of caution is that elevation in serum DHEA has been associated with male pattern baldness in young men[76]. DHEA has not been associated with increased risk for prostate cancer[77]. However, in men with very low testosterone levels, conversion of DHEA to testosterone could theoretically increase the risk of prostate cancer. Thus, it would be prudent for men to check with their physicians before initiating DHEA therapy. Although DHEA does not appear to pose significant risk for breast cancer in young women with high estrogen levels, it may be more of a risk for postmenopausal women whose estrogen levels might be increased by intake of high doses of DHEA[78]. Thus, postmenopausal women should not take DHEA without the guidance of their physician.

FISH OIL (*OMEGA-3* FATTY ACIDS, EPA, DHA)

The human body can synthesize saturated fatty acids from carbohydrate, and it can produce monounsaturated fatty acids from saturated fats. However, there are two types of fatty acids that the body cannot synthesize, the polyunsaturated *omega-3* and *omega-6* fatty acids. These are the so-called "essential" fatty acids. We must obtain them from our diet to maintain health.

The only truly essential *omega-3* fatty acid is alpha-linolenic acid. The human body is able to use alpha-linolenic acid as substrate to synthesize two other extremely important *omega-3* fatty acids, eicosapentaenoic acid (EPA) and docosahexaenoic acid (DHA). The advantage of fish oil, which is rich in EPA and DHA, over flax seed oil, which is rich in *alpha*-linolenic acid, is simple to explain. Although the body can synthesize EPA and DHA from *alpha*-linolenic acid, it is not terribly efficient in doing so. It is better to consume EPA and DHA in fatty ocean fish or in supplements than to rely on their production from *alpha*-linolenic acid in vegetable oils.

While both types of fatty acids are required for health, the *omega-3* and *omega-6* are clearly different from one another. They each have their role, and their intake needs to be balanced. The ratio of *omega-3* to *omega-6* polyunsaturated fatty acids in our diet has changed through the centuries. The polyunsaturated fats from the vegetable oils we commonly consume from salads and cooked foods are unbalanced in terms of containing too much *omega-6*. Thus, we modern humans are faced not only with providing ourselves adequate amounts of the essential fatty acids, but also obtaining them in proper proportion.

Aside from their roles as energy storage molecules and building blocks of cell membranes, the polyunsaturated fatty acids are necessary to produce a number of critical chemical messengers in the body. The powerful, hormone-like eicosanoids are produced from metabolites of the *omega-3* fatty acid alpha-linolenic acid and the *omega-6* fatty acid arachidonic acid. However, the resulting eicosanoids are quite different from each other. For example, the eicosanoids derived from *omega-6* fatty acids tend to have inflammatory effects, whereas those derived from *omega-3* fatty acids tend to dampen inflammation.

One of the most consistent set of findings in the medical literature is that an adequate dietary intake of *omega-3* fatty acids, such as are found in fatty ocean fish, is an important part of the prevention and treatment of Metabolic Syndrome. Diets high in fish help maintain normal glucose tolerance[79], and they may offer a certain degree of protection from the development of diabetes type II[80]. There also tends to be reduction in serum triglycerides among communities where fish is a staple in the diet[81]. Not surprisingly, epidemiological studies have shown that diets rich in fatty fish reduce the risk of cardiovascular disease[82], which is a common end point of Metabolic Syndrome.

A wealth of studies shows that supplementation of *omega-3* fatty acids, particularly in the form of fish oil, can improve various aspects of Metabolic Syndrome. Fish oil is well known to reduce both serum triglycerides[83] and postprandial lipemia[84]. It increases both HDL and LDL cholesterol, which has led some physicians to cast a wary eye upon its use. However, while increasing levels of total LDL, it reduces the proportion of LDL that is small, dense and most atherogenic[85]. *Omega-3* fatty acids help relieve hypertension and improve platelet function[86]. Supplementation with fish oil can help maintain insulin sensitivity in normal subjects[87]. This maintenance of insulin sensitivity may be due in part to the ability of *omega-3* fatty acids to stimulate PPAR in skeletal muscle, and enhance the mobilization and partitioning of fatty acids taken up into muscle cells. Unfortunately, *omega-3* fatty acids do not appear to restore insulin sensitivity once diabetes type II has emerged.

Omega-3 fatty acids also help reduce many of the adverse, secondary effects of Metabolic Syndrome. Fish oil and *omega-3* fatty acids dampen the systemic inflammation seen in Metabolic Syndrome[88]. They may do so by providing the substrate for the synthesis of anti-inflammatory prostaglandins[89]. These fatty acids reduce serum levels of CRP and inflammatory cytokines, such as TNF-alpha[90]. They also block activation of one of the primary triggers of inflammation, NFk-B, through interaction with PPAR[91]. There has been concern over the fact that *omega-3* fatty acids are easily oxidized, and may contribute to lipid peroxidation. In this regard it should be noted that it is the oxidized form of *omega-3* fatty acid that most readily activates PPAR.

Fish oil, and the EPA and DHA it contains, helps prevent Metabolic Syndrome and ameliorate the adverse effects this condition produces. Because DHA and EPA each have slightly different salutary effects on Metabolic Syndrome, it seems most reasonable to supplement the diet by addition of fish oil, which contains both[92].

There are many ways in which the *omega-3* fatty acids EPA and DHA impact brain function. Of particular significance is the fact that large amounts of these fatty acids, particularly DHA, are found in the human brain. Adequate supply of DHA is necessary for normal development of the fetal and infant brain. Infants deprived of adequate DHA show poor performance of cognitive and psychomotor skills[93]. DHA is integrated into neuronal membranes in the form of various phospholipids, and it has an important role in maintaining the structural integrity and function of neurons in the human brain[94]. Animal studies have shown that diets depleted in *omega-3* fatty acid result in abnormalities in the activities of monoamine neurotransmitter systems in the brain. Researchers found increases in serotonin 2A and decreases in dopamine 2 receptors in the frontal cortices of *omega-3* fatty acid deficient rats[95]. In humans, such changes might be expected to contribute to Major Depression. Like lithium and other mood stabilizers, *omega-3* fatty acids have mild ability to inhibit the activity of protein kinase C[96] and stimulate phosphatidylinositol 3-kinase[97]. The anti-inflammatory effects of these fatty acids might also be expected to counteract some of the adverse inflammatory effects of Metabolic Syndrome in the brain mediated by cytokines such as TNF-alpha, and triggers of inflammation such as NFk-B. There are ample mechanisms by which the *omega-3* fatty acids might protect against Major Depression, Bipolar Affective Disorder and schizophrenia. However, while epidemiological studies of diet and psychiatric illness are consistent with such beneficial effects, studies of the use of fish oil and other *omega-3* fatty acid preparations to treat psychiatric illnesses have been rather disappointing.

Several epidemiological studies have suggested that low consumption of fish rich in *omega-3* fatty acids increases the risk of depression[98]. However, not all studies have shown such a strong relationship. In one case, the relationship was seen only in women[99]. In other cases, no compelling relationship was found at all[100]. Studies are more consistent in showing that low serum levels of *omega-3* fatty acids, particularly DHA[101], are associated with depression[102]. There are also reports of reduced levels of DHA in the adipose tissue of depressed patients[103]. Together, studies tend to suggest that depletion of body stores of *omega-3* fatty acids, whether due to poor intake or other causes, may contribute to Major Depression.

Studies of the treatment of depression with specific *omega-3* fatty acids or fish oil rich in EPA and DHA have shown rather weak and inconsistent effects. To some degree, this may be due to the fact that so many different methods have been used to evaluate the effects of *omega-3* fatty acids. Addition of fish oil to antidepressants was found to be helpful in one study[104], but not in another[105]. Monotherapy of Major Depression with DHA is ineffective[106]. However, addition of the EPA derivative ethyl-EPA to antidepressants was found to be quite helpful in patients with residual symptoms and in a group considered to be genuinely resistant to standard treatment[107].

Inconsistencies are also seen in reports of potential benefits of fish oil in the prevention and treatment of postpartum depression[108], and in the depressed mood states associated with conditions as varied as chronic fatigue syndrome, borderline personality disorder, and obsessive compulsive disorder[109]. Although fish oil and specific *omega-3* fatty acids may be helpful, there is no consistent or compelling evidence that they have a critical role in the treatment of Major Depression.

An epidemiological study has shown a strong correlation between the consumption of seafoods rich in *omega-3* fatty acids and reduction in the prevalence of BPAD in its various forms[110]. Moreover, there are indications of abnormally low levels of DHA in erythrocyte membranes in patients with bipolar mania[111]. However, as in Major Depression, studies of the usefulness of *omega-3* fatty acids in the treatment of the depression of BPAD have yielded inconsistent results. In some cases, *omega-3* fatty acids have been found to be helpful[112], whereas in others they were determined to be ineffective[113]. Whereas some researchers suggest that fish oil calms the irritability of mania and hypomania[114], others note that these fatty acids may be more useful in bipolar depression than in mania[115]. Even the most recent reviews of the use of fish oil and *omega-3* fatty acids for the treatment of the various manifestations of BPAD have described the results as "conflicting"[116]. I must note my personal impression that the literature contains more reviews on the subject than actual studies!

The oldest and largest literature regarding the role of *omega-3* fatty acids in a psychiatric disorder is that concerning schizophrenia. The first suspicions of a relationship between *omega-3* fatty acid metabolism and psychosis was voiced as early as 1981[117]. Epidemiological studies have since shown that a more favorable course of illness occurs with a higher consumption of polyunsaturated fats from vegetables and seafood than from the sources of saturated fat such as land animals and birds[118]. *Omega-3* fatty acid content of erythrocyte membranes is low in patients with schizophrenia[119]. This has even been found to be the case in individuals suffering their first episode of psychoses[120], which suggests that these deficits exist early in the illness.

Again, while epidemiological data and studies of serum fatty acid levels strongly suggest a role for *omega-3* fatty acids in schizophrenia, results of many studies using these substances in the treatment of schizophrenia simply do not bear out this suspicion. A 2006 review written for the well-regarded Cochrane Database sees the results of years of studies on potential benefits of fish oil, EPA and DHA on schizophrenia as still "inconclusive"[121]. Their most encouraging conclusion was that supplementation with EPA may allow some reduction in the amount of neuroleptic required to control symptoms of schizophrenia.

Diet is an important factor in determining risk for developing Alzheimer's Disease. Total fat, particularly saturated animal fat, disposes individuals to Alzheimer's Disease due in part to its tendency to enhance inflammatory processes. Lipid oxidation, which occurs in part as a result of inflammatory processes, appears to be an early and common occurance in Alzheimer's Disease[122].

Consumption of fish containing high levels of *omega-3* fatty acid appears to counteract some of that increased risk for Alzheimer's Disease. This is likely due to the ability of the *omega-3* fatty acids to dampen inflammation[123]. People who consume fish at least once a week can have a 60 percent reduction in the likelihood of developing Alzheimer's Disease. This reduced risk of Alzheimer's Disease is attributed primarily to an increased intake of DHA[124]. Data from the well-known Framingham Heart Study has been investigated and similarly determined to reveal that high serum levels of DHA are associated with reduced risk of developing Alzheimer's Disease[125]. Interestingly, the protective effect of fish consumption is not seen in individuals who carry the APOE epsilon 4 allele that is known to predispose carriers to Alzheimer's Disease[126].

Serum levels of DHA and EPA are often found to be lower in Alzheimer's Disease patients than in control subjects. Levels of DHA in particular have been found to exist in reverse proportion to the severity of Alzheimer's dementia[127]. The brains of Alzheimer's Disease patients also have low levels of DHA and other long-chain polyunsaturated fatty acids while having disproportionately high levels of saturated fats. These changes are related to Alzheimer's Disease and not simply to ageing alone[128].

Supplementation of Alzheimer's patients with fish oil[129] or other *omega-3* fatty acid preparations, such as ethyl-EPA[130], have been disappointing. This is possibly because by the time the illness is diagnosed, the damage is already done. Benefits of *omega-3* fatty acids in Alzheimer's Disease have been reported, but this tends to be only in the most mild cases of the illness. Overall it appears that fish oil is more useful in preventing than in treating Alzheimer's Disease.

Curiously, both arrythmogenic and antiarrythmic[131] effects have been reported with the use of *omega-3* fatty acids. Recent reviews on this subject have suggested that the pro-versus antiarrythmic effects may depend upon the underlying cause of the predisposition to arrhythmia. Whereas *omega-3* fatty acids may reduce the risk of arrhythmia in post myocardial infaction, it may increase the likelihood in an individual with active ischemia or preexisting reentry arrythmias[132]. Overall, the most common report is that high intake of fish oil is inversely related to the risk of sudden cardiac death[133]. If an individual has significant cardiac disease or an existing reentry arrythmia, it would be prudent to consult his or her cardiologist prior to suggesting the addition of *omega-3* fatty acids. There are also some concerns voiced about increases in lipid peroxidation with high intake of *omega-3* fatty acids[134]. However, while a theoretical concern, there is little evidence of this actually being any sort of a risk factor among patients. Moreover, it becomes still less of a concern if adequate amounts of fat soluble antioxidants, such as vitamin E, vitamin A, and alpha-lipoic acid, are consumed in the diet or as supplements. There is also some legitimate concern about consuming heavy metals in fish or fish oil preparations. Overall, the data suggests that, with exception of certain fish that accumulate heavy metals, that is swordfish and shark accumulating mercury, the benefits of eating fish is worth the small risk[135]. Many good, over-the-counter preparations of fish oil have negligible amounts of mercury and other heavy metals[136].

Epidemiological studies have generally shown that low intake of fish or other sources of *omega-3* fatty acids is associated with increased risk of Major Depression, BPAD, and Alzheimer's Disease as well as poor outcome in schizophrenia. This data is consistent with reports of low levels of these fatty acids in erythrocyte membranes and, in some cases, adipose tissues. There are also many physiological mechanisms by which the *omega-3* fatty acids could have a salutory effect upon brain function and mental illnesses. Part of the benefit could be reduction of the risk of Metabolic Syndrome and dampening of the adverse manifestations of the syndrome on brain tissue. Nonetheless, while all these factors make the use of fish oil seem an extremely attractive course of action, results of studies of the use of this substance in the actual treatment of psychiatric disorders have been inconclusive and rather disappointing.

The most reasonable explanation for the apparent lack of efficacy of *omega-3* fatty acids in the treatment of psychiatric illness is that, while adequate levels may be necessary to maintain mental health, once the symptoms emerge, it may be too little and too late to reverse the symptoms by reintroducing the fatty acids. It may also be that the benefits of the reintroduction of *omega-3* fatty acid in the diet occur over longer periods of time than the weeks or months of treatment that have been observed under experimental conditions. It is possible that these fatty acids play a permissive role in recovery from psychiatric illness. That is, while not sufficient, they may be necessary over the long run to allow full remission from illness. These possibilities remain to be fully investigated. Nonetheless, in view of evidence that *omega-3* fatty acids can help some individuals, and a lack of evidence of any significant adverse effects, it seems prudent to take advantage of the possibility that these fatty acids may be of help. I personally advise my patients to take a gram of fish oil three times a day with meals as part of my overall treatment of their psychiatric illness.

GINSENG

Ginseng has been used as medicine in Asia for thousands of years. It has generally been seen as a restorative, tonic medicine for the weak and elderly. However, it has also been viewed as an herb that can help even healthy individuals resist stress. There are a variety of ginseng plants from around the world. The most common forms are Asian Ginseng and American Ginseng. They are similar in their effects, although Asian Ginseng is often said to be more active than the American plant. So-called Siberian Ginseng is a closely related plant with many similarities to the true ginsengs. However, most studies of relevance to Metabolic Syndrome have been performed with Asian or American Ginseng.

As is often the case with herbal medicines, the extracts of ginseng are extremely complicated in makeup. They always contain a variety of related compounds, generally referred to as ginsenosides or saponins. The different extraction processes, sources and types of ginseng can result in significant differences in the concentrations or even the very presence of important medicinal components of the ginseng. Standardized extracts from reputable herbal medicine sources should be used.

There is an extensive literature showing that ginseng increases insulin sensitivity and reduces postprandial glucose in animals. It has also been found to reduce inflammatory processes, such as levels of TNF-alpha in laboratory rats. There is now a growing literature showing that ginseng may have similar effects in humans. Chronic use of ginseng enhances insulin sensitivity in patients with diabetes type II. It reduces serum levels of insulin and glucose, both in the fasting state and during glucose tolerance tests[137]. It even improves psychomotor skills and elevates mood in such patients[138]. Ginseng reduces anxiety and fatigue, as well as decreases levels of cortisol, a major contributor to Metabolic Syndrome[139].

The means by which ginseng reverses some of the signs of Metabolic Syndrome are not clear. However, application of ginseng extract to human cells causes effects similar to those of insulin. It increases uptake of glucose into sheep cells[140], again suggesting that it mimics or enhances insulin's effects. Ginseng also stimulates release of insulin from the pancreas[141]. This occurs during the so-called first phase of release, the lack of which may contribute to a hypoglycemic effect when the pancreas tries to catch up with a large, second phase release of insulin. This is the phenomenon that Dr. Seale Harris referred to as dysinsulinism. It is a harbinger of the transition from Metabolic Syndrome into outright diabetes.

Some components of ginseng stimulate the cellular factor PPARgamma[142]. This is the factor stimulated by many of the newest and most effective diabetic medications. In fact, one component of ginseng is as potent as the drug troglitazone in stimulating PPAR. PPARgamma increases storage of fat, and thereby reduces dangerous levels of free fatty acids and triglycerides in the blood. Ginseng also stimulates production of type IV glucose transport molecules in cell membranes. This is one of the major effects of insulin, and one of the primary ways in which insulin lowers postprandial serum glucose back to fasting levels.

Ginsenosides inhibit the pro-inflammatory nuclear signal by NFk-B in mice[143]. This in turn would be expected to reduce high levels of inflammatory cytokines such as those seen in Metabolic Syndrome. Ginseng reduces production of one of the major inflammatory substances, TNF-alpha, in human macrophages[144]. Macrophages participate in the inflammatory processes that lead to atherosclerosis in the arteries of the heart and other areas of the body.

Ginsenosides share the molecular structure of steroid hormones. Indeed, certain ginsenosides can mimic the effects of estrogen[145]. There is even some concern about the use of ginseng by women with, or at high risk for, breast cancer. In animal models of Alzheimer's Disease, ginsenosides act in a fashion similar to estrogen in reducing the brain inflammation that causes further progression of the illness[146]. Ginsenosides also interact with receptors for cortisol. Because these molecules are fat soluble, they can affect cortisol receptors in the brain.

In the brains of old rats, ginseng can change the balance between mineralocorticoid and glucocorticoid receptor activity in the brain. This balancing of the activity of the types of cortisol receptors may have antistress and antidepressant effects[147]. However, it is unknown if human brains are affected in the same ways observed in rat brain.

Ginseng has been considered as a tonic, restorative substance for the brain and the body. In animal studies, ginsenosides have reduced inflammation in the brain, and reduced amounts of beta amyloid in brain tissue[148]. However, in the few studies in which ginseng was tested as a means to improve cognitive function in the elderly and demented, it has generally failed to give discernible benefit[149]. In younger subjects, the addition of ginseng to a regimen of vitamins did seem to give extra benefits in general quality of life factors, particularly in groups of subjects suffering the "stress of high physical and mental activity"[150].

Overall, ginseng is a benign substance with no major adverse effects. Daily doses of 100–200 mg of standardized extract are generally very well-tolerated. The only caveat is that women with a history of breast cancer seek medical counsel before starting treatment with ginseng. The major question is whether ginseng, which can be rather expensive, offers enough benefit to make it worth buying and taking on a regular basis.

LEUCINE

Recently, there have been some fascinating scientific reports on the role of leucine in appetite control and hormone regulation. Leucine is a unique substance in the body. It is one of only nine essential amino acids. It is one of only three branched chain amino acids, and the only essential one in that group. It is also the only amino acid that cannot be metabolized into glucose. When the body has a unique molecule, it is not uncommon for that molecule to be used for special purposes. It is becoming apparent that the body may use leucine as a signaling molecule that helps the body sense its protein and energy status.

As molecular biology becomes more a part of medical science, some fascinating terms and phrases from that area of science are popping up in medical parlance. One example is mTOR, which stands for the "mammalian target of rapamycin". Rapamycin is an antibiotic produced by the *Streptomyces hygroscopicus* bacteria. It decreases certain components of the human immune response, and it is useful for preventing rejection of organs after transplant surgery. A major site of rapamycin actvitiy is mTOR. The mTOR site is deeply imbedded in the chemistry of human cells. It seems to be involved in controlling cell growth and protein synthesis. It also mediates some of the effects of insulin in cells. It has been known for several years that leucine stimulates mTOR in the brain and other parts of the body.

In a study recently published in the journal *Science*, injection of leucine directly into the hypothalamus of the brains of rats resulted in substantial reductions in their food intake[151]. Apparently, the injection of leucine into the brain reduces appetite by stimulating the mTOR site. When rats were administered rapamycin, which inhibits the activity of mTOR, they showed an increase in appetite and food intake. Injection of small amounts of leptin into the brain also stimulates mTOR and decreases appetite. Blocking the activation of mTOR with rapamycin blunts the ability of leptin to reduce food intake. Leucine also appears to play a role in the production and release of leptin from fat cells. This effect of leucine again appears to be mediated by stimulation of mTOR. In test tube studies of fat cells from rats, addition of leucine was found to increase their release of leptin. This effect of leucine was blocked by the mTOR antagonist rapamycin[152]. When leucine is removed from the diets of rats, the serum levels of leptin after meals are found to decrease by nearly 40 percent[153]. These data show a strong relationships between leucine, leptin, mTOR and control of appetite.

As an essential amino acid, leucine is required in the diet for production of various proteins in the body. It has been calculated that the body needs at least 2 g of leucine a day to fulfill this role. However, leucine does many other things besides serving as a building block in the synthesis of protein. Dr. Donald Layman of the University of Illinois at Urbana-Champaigne is an authority on the role of leucine and other amino acids in energy metabolism and weight control. He believes that because of all the jobs performed by leucine in the body above and beyond its use in protein synthesis, as much as 8 grams a day might be necessary to take full advantage of its effects[154]. For example, in muscle cells, leucine serves as a gauge for the availability of all the other amino acids required for protein synthesis. When high concentrations of leucine are present, it enhances the effect of insulin in stimulating protein synthesis in muscle. This effect of leucine allows muscle mass to be maintained during periods of weight loss, as long as sufficient protein is consumed. Interestingly, leucine enhances protein synthesis in muscle cells by stimulating mTOR. Leucine also serves an important role in maintaining serum glucose levels during fasting states. As leucine is broken down in muscle cells for use as fuel, it donates an amino group to the substance pyruvate, which results in the amino acid alanine. Alanine returns to the liver, dumps off the amino group and gets converted into glucose for organs like the brain to consume. Thus, sufficient leucine is a necessary part of any successful high protein diet.

It is sobering to note that along with its use as an immuosuppressant, rapamycin is beginning to be used by doctors to treat cancer. In fact, a number of substances that decrease activity of mTOR may have potential as cancer treatments[155]. This would lead any prudent doctor to wonder if overstimulation of mTOR by indulging in supplementation with leucine might put susceptible individuals at risk for cancer. My concern over this possibility led me to consult some authorites on the subject of cancer, leucine and human physiology. For the most part, the experts were not terribly concerned about increases in leucine causing or aggravating cancer. Christopher Lynch, Ph.D. a professor of Cellular and Molecular Physiology at the Pennsylvania State College of Medicine informed me that the body has "enormous capacity" to metabolize extra leucine, and the likelihood of producing sustained increases in serum leucine by any reasonable method was rather remote. He doubted that high intake of leucine would result in increased risk for cancer. He described animal experiments in which genetic engineering reduced the ability to metabolize leucine. That, in turn, resulted in roughly 20-fold increases in serum leucine. He noted no increases in development of tumors in those animals. He further explained that abnormalities "upstream" from mTOR may cause persistent activation, which, in turn, may contribute to malignancy. In other words, stimulation of mTOR by leucine is probably not what rapamycin is reversing when it is used to fight cancer. Finally he states that the body is likely to quickly adapt to overstimulation of mTOR by leucine, through a process of down-regulation. That is, mTOR will simply no longer be as responsive to leucine stimulation.

Vickie Baracos, Ph.D., is a Professor in the Department of Oncology at the University of Alberta in Edmonton, and the Alberta Cancer Foundation Chair of Palliative Medicine. Dr. Baracos agreed that the body is quite adept at dealing with extra amino acid in the body, including additional leucine. She informs me that the number of the enzymes that break down leucine would be increased in the presence of extra leucine. Consequently, she describes herself as hard-pressed to imagine any convenient way to to obtain consistently high levels of leucine in the body. She finally felt compelled to inform me that purified amino acids such as leucine taste awful! All in all, these experts gave little indication that increasing leucine intake would pose any danger to people. On the other hand, neither thought that increasing leucine intake would produce dramatic increases in leucine levels or any secondary effects upon appetite. In that regard, leucine supplements might simply be a waste of money.

Donald Layman, Ph.D., a professor in Nutrition at the University of Illinois was also skeptical about leucine supplementation posing any risks for cancer. He did note, however, that a recent report indicates that under certain circumstances, such as severe burns, leucine stimulation can override down-regulation of mTOR[156]. Thus, while not seriously concerned, he noted the remote possibility that leucine could escape the usual control mechanisms and cause problems. Nonetheless, in this context it is worth noting that Dr. Layman is studying the use of leucine in treating the severe loss of weight and lean body mass that can occur in cancer patients.

Another possible "dark side" of leucine supplementation is a theoretical potential to exacerbate depression. Although the amino acids differ from one another in the degree in which they are soluble in water and fat, none can simply pass through the fatty blood-brain barrier to reach the brain tissue. They are all carried across the blood-brain barrier by special amino acid transport molecules. One type of amino acid transporter carries what are called the large neutrally charged amino acids. This group includes leucine, valine, leucine, isoleucine, tyrosine, phenylalanine and tryptophan. A problem may arise when these amino acids compete with one another for sites on the transporter molecule to be carried into the brain. If there is an overabundance of leucine, this could result in the others being pushed out of transport spots which would, in turn, produce relative depletion of tryptophan, tyrosine and phenylalanine in the brain[157]. Those amino acids are used in the brain to produce the neurotransmitters serotonin, noradrenaline, and dopamine. Depletion of those substances in the brain, particularly serotonin, is known to exacerbate depression[158].

A final concern about leucine supplementation arises from the existence of several rare but serious genetic defects in the metabolism of leucine. Some individuals are born without the ability to break down leucine. Infants born with this defect will die if not treated promptly. Treatment consists of restriction in the intake of leucine and the other branched-chain amino acids. Intake is limited to only what is absolutely essential for protein synthesis. Because of the large amount of branched-chain amino acids secreted by the kidneys, the urine of these children takes on a peculiar, sweet smell similar to maple syrup. Hence the disease has come to be known as Maple Syrup Urine Disease.

Although the most severe form of Maple Syrup Urine Disease can be fatal in infancy, there are less severe forms that cause problems only during severe illness or other types of stress upon the body. It seems unlikely that an individual would be unaware of having this condition. Nonetheless, such individuals would certainly be ill-advised to supplement their diet with extra leucine. It is worth stating that Susan Hutson, a biochemist at Wake Forest University Sachool of Medicine, has noted the ability to metabolize leucine in humans as being far less than that in rodents[159]. Thus it is possible that if assumptions about the ability of humans to dispose of leucine are being drawn from animal studies, that ability may be overestimated.

Existing data lead me to be somewhat hesitant to recommend supplementation with leucine. Although dangers of too much leucine cannot be dismissed, the more likely danger is one of throwing your money away. Thus, the most prudent approach may be to exploit the benefits of leucine by increasing intake of good dietary sources of the amino acid. Some of the best sources of leucine are dairy products. There are many reports suggesting that dairy products are useful in diets for weight loss. In fact, it has been suggested that the relatively high concentration of leucine in milk, which is roughly 12 percent of the total protein versus 8 percent in meat, may be at least partially responsible for weight loss observed in individuals that increase their consumption of dairy products[160]. In any case, multiple benefits could be gained by this natural approach.

NIACIN (NICOTINIC ACID, VITAMIN B3)

Niacin is best known as a member of the B-complex family of vitamins. It began with the name nicotinic acid, as it was first synthesized in 1867 through oxidation of nicotine. However, its place in physiology was not recognized at that time. The role of niacin as a vitamin with an ability to cure pellagra was discovered by the biochemist Conrad Elvehjem in 1937. After it was found to be a life-sustaining vitamin, the name nicotinic acid was viewed as unseemly. The name niacin was chosen, and was said to have been derived from ni-cotinic ac-id and vitam-in.

Niacin is transformed into its primary active form nicotinamide adenine dinucleotide (NAD). It is involved in a multitude of fundamental physiological processes including fat and carbohydrate metabolism. NAD is essential for the synthesis of steroid hormones and, ironically, cholesterol. Although it is a vitamin, the human body has the ability to synthesize niacin out of the amino acid tryptophan. If tryptophan intake is sufficient, the body can manufacture an adequate amount of niacin. Unfortunately, niacin deficiency can occur, and it results in the wretched illness, pellagra. Pellegra is characterized by dementia, depression, dermatitis and diarrhea. As recently as the 1920s pellagra ranked eighth among the most common causes of death in the southern United States[161]. After addition of niacin and other vitamins to food began in the late 1930s, the incidence of pellagra in the United States has become vanishingly small.

The remarkable ability of niacin to reduce LDL, VLDL and triglycerides, while increasing HDL was discovered in 1955 by the Canadian pathologist Rudolf Altschul[162]. Doses of 2–6 g of niacin can decrease serum triglycerides by up to 80 percent and LDL by 10–15 percent. HDL levels are increased 15–30 percent[163]. Curiously, the increase in HDL is somewhat less in patients that start with low HDL levels than in those with normal HDL[164]. It almost seems unfair.

It is not entirely clear how niacin acts to fight dyslipidemia. One mechanism is likely the inhibition of lipolysis in fat cells, which in turn reduces the amount of free fatty acids that can be delivered to the liver. This results in a decrease in the production of triglycerides by the liver and, consequently, a decrease in the production and release of VLDL that is used to package and transfer the triglyceride[165]. The reduction of VLDL is likely to reduce the attrition rate of HDL, but the exact mechanism is unknown.

What is important to understand is that the hypolipidemic effects of niacin do not arise out of its role as a vitamin. The recommended daily allowance of niacin is 18 mg a day. There are nutritionists who believe that as much as 115 mg a day of niacin may be an "ideal" daily intake[166]. Nonetheless, such doses are far below the multigram doses required to affect serum lipids. Moreover, it has recently been found that niacin can bind to a unique receptor that is coupled to a G-protein[167]. This receptor, known as HM74A, may modulate the effects of high doses of niacin on lipids. Niacin may also stimulate PPAR in certain cell types through activation of this receptor[168]. It is not yet known what the endogenous ligand for this receptor might be.

At the doses necessary to reverse dyslipidemia, niacin can have toxic effects. One effect, which is more uncomfortable than dangerous, is the skin flushing response[169]. It produces a feeling not unlike a prickly sunburn. Aspirin can sometimes diminish the flush response, and this is due to the fact that the response is mediated in part by the synthesis of prostaglandins. This effect lessens with time. Niacin can also cause gastrointestinal problems, and individuals with a history of ulcers should steer clear of it. Because niacin raises uric acid, patients with gout should not be started on this medicine.

The most dangerous side effect that can be seen with niacin treatment is hepatotoxicity[170]. The risk is increased by the use of sustained release forms of niacin or combining niacin with other hypolipidemic drugs, such as statins[171]. However, rather than contraindicating the use of niacin in monotherapy or combination therapy, it simply illustrates the fact that this "vitamin" is not innocuous and must be monitored like any other medication.

Another concern about niacin has been that it can increase serum glucose and decrease insulin sensitivity[172]. For a time there was reluctance to use niacin to treat hyperlipidemia in diabetic patients. One might have similar concerns in respect to treating patients with Metabolic Syndrome. However, the consensus has been that the benefits of niacin far outweigh the risks. A paper from the Coronary Drug Project supported the use of niacin in post-myocardial infarction patients with or without Metabolic Syndrome. In fact, among various treatments for hyperlipidemia, niacin was the only treatment that significantly reduced recurrence of MI and decreased mortality rates[173]. A European Consensus Panel for the treatment of Metabolic Syndrome has published a position paper in which they recommend adding niacin to a statin for the trteatment of hyperlipidemia in patients with Metabolic Syndrome and Diabetes[174].

It is of at least historical interest to point out the fact that one of the coauthors of Altschul's landmark report on niacin and cholesterol was Dr. Abram Hoffer. Dr. Hoffer, along with Dr. Humphrey Osmond, was one of the founders of Orthomolecular Psychiatry. For better or worse, he is probably best known for promoting high, so-called "mega doses" of niacin for the treatment of schizophrenia. This treatment was never embraced by mainstream psychiatry, but rather was refuted in several papers[175]. It is worth noting that some basis for Hoffer's theory about adrenochrome being an endogenous neurotoxin responsible for schizophrenia has recently been established. That is, there is compelling evidence of defects in the enzyme glutathione S-transferase in some schizophrenics. One function of this enzyme may be to rid brain tissue of adrenochrome and other toxic oxidation products of catecholamines[176].

Another remarkable finding has been that patients with schizophrenia tend to have diminished skin flushing responses to niacin. This apparent fact was exploited by the late Daniel Horrobin as a diagnostic test for schizophrenia[177]. Although the validity of the test has been questioned over the years, some recent reports provide support for the original observation[178]. Because of mediation by prostaglandins, this diminished response to niacin has further led to a hypothesis that a fundamental biochemical defect in schizophrenia may be a relative inability to synthesize certain prostaglandins. This could be due to deficits in the enzymatic machinery, or simply to a low reserve of the raw material from which prostaglandins are produced. Defects in *omega-3* and *omega-6* fatty acids, which are precursors for prostaglandins, have been observed in schizophrenia[179].

RESVERATROL

It has been called "The French Paradox". Despite their penchant for pastries, gravies, and rich, high-calorie foods, the French have a surprisingly low rate of heart disease. This apparent immunity of the French from the degenerative diseases the rest of western civilization are prone to may have first been noticed by an Irish physician, Dr. Samuel Black as far back as 1819. He attributed this to, "the French habits and modes of living, coinciding with the benignity of their climate and the peculiar character of their moral affections". The bottle of red wine, ever present on the French dinner table, may have something to do with their resistance to the effects of imprudent diet. The question for scientists has been, "what is it in red wine that could provide health benefits for people with high risk for Metabolic Syndrome and cardiovascular disease?"

There is compelling evidence that moderate consumption of alcohol can offer some protection from Metabolic Syndrome, diabetes and heart disease. However, wine is a complicated mix of alcohol and a variety of plant-derived substances that give the wine color and flavor. Anyone of those substances could be providing additional health benefits. One substance in red wine that has been suspected of contributing to its beneficial effects on health is resveratrol. Resveratrol is a multiringed molecule, a "polyphenol", that is produced in grapes and some other plants in response to fungus infections and other environmental stresses. Peanuts, mulberries, cranberries, lilies, and eucalyptus are among the many and various types of plants that also produce resveratrol. Similar molecules can be found in other types of plants that serve functions like those of resveratrol in protecting the plants from stresses.

Many remarkable effects have been attributed to resveratrol, ranging from benefits in treating heart disease and cancer, all the way to lengthening life itself. Most of these effects have been observed in studies using laboratory animals. Moreover, many of those studies used doses of resveratrol much higher than those that can be obtained from drinking wine. While such studies do not address the question of whether resveratrol is responsible for "The French Paradox", they still could provide the basis to use high doses of resveratrol as a supplement. A number of studies have evaluated effects of resveratrol in human tissues, but these have most often been *in vitro*. Still, many of the results of resveratrol studies have been dramatic enough to make any medical scientist take notice of them.

In various animal studies, resveratrol has reduced deposition of fat in the lining of arteries[180], reduced hypertension by enhancing the effects of endothelial nitric oxide synthase[181], mimicked effects of insulin[182], and lessened the toxic effects of amyloid protein in brain tissue. That is the same amyloid that is so destructive in Alzheimer's Disease[183]. In studies in which the effects were evaluated in blood or tissue taken from humans, resveratrol was found to reduce the inflammatory effects of TNF-alpha on the lining of blood vessels[184], and prevent the oxidation of LDL that enhances deposition of fat and cholesterol in the arteries[185].

One of the most interesting effects of resveratrol has been its ability to stimulate a cell factor known as SIR-2. When animals are greatly restricted in their calorie intake, they are often found to live longer. Calorie restriction has been found to stimulate SIR-2, which is at least partially responsible for the increase in longevity. Resveratrol has been found to enhance the activity of SIR-2 in yeast, and to extend the life-span of those primitive organisms. *In vitro*, resveratrol stimulates the activity of a human analog of SIR-2, known as SIRT-1[186]. Resveratrol extends the life span of the primitive worm *Caenorhabditis elegans*, a commonly used laboratory invertebrate. Resveratrol also extends the lives of fruit flies as well as those of the small vertebrate fish, *Nothobranchius furzeri*[187].

Unfortunately, there is a downside to resveratrol research. A major argument against resveratrol being an imporatnt component of red wine's effects, or even an effective treatment in its own right, is the fact that it has extremely low bioavailability. Bioavailability is a term that means pretty much what it says, that is, it is a measure of how much of a substance is actually available inside an animal to bring about effects. Some drugs are very poorly absorbed, and thus are limited in bioavailability from the very beginning. About 70 percent of ingested resveratrol gets into the human bloodstream, which is a reasonably high absorbtion rate. However, little if any of the pure resveratrol remains after what is absorbed from the intestine makes its first pass through the liver[188].

The liver changes almost all of the resveratrol into sulfates and glucuronides, which the body sees as deactivated metabolites marked for disposal.

Resveratrol could remain in high enough concentration in certain areas of the body, such as the intestine, to produce therapeutic effects, including its alleged anticancer effects. Unfortunately, existing studies make it difficult to understand how even fairly high doses of resveratrol could do anything else of significance if ingested by real, living human beings. The only study even mildly contrary to that conclusion is a study in which pre- and postmenopausal women were asked to consume 36 g a day of freeze-dried grape skin powder[189]. This powder was found to be rich in resveratrol, and it improved serum lipid profile and eased oxidative stress. However, the preparation also contained large amounts of other substances, such as quercetin, myricetin, anthocyanins, and others that, like resveratrol, are known to have antioxidant and other properties. It would have been difficult to attribute any particular effects specifically to resveratrol. In any case, effects beneficial for treatment and prevention of Metabolic Syndrome included reductions in plasma triglycerides and LDL cholesterol. There were also signs of reductions in oxidative stress and the inflammatory processes it stimulates.

There is little evidence that resveratrol is toxic, but a significant concern about resveratrol is that it can mimic the effects of estradiol[190]. Indeed, its molecular structure is quite similar to that of the synthetic estrogen, diethyl-stilbesterol. The concern is that resveratrol could stimulate growth of breast cancer in certain susceptible women. Very small amounts of resveratrol are ingested when drinking red wine, and this raises little concern. However, if resveratrol were purified and taken in large amounts for its alleged health benefits, the risks of breast cancer could become significant.

Continue drinking moderate amounts of red wine if you enjoy it, as there is substantial evidence that it can help prevent or at least slow the progression of Metabolic Syndrome. If the small amounts of resveratrol and other compounds from the grape skins in it are adding to your health, so much the better. Hopefully, a time will come when a clever, pharmacological method can be devised to deliver unaltered resveratrol to tissues in the body that can benefit from it. At present, resveratrol as a supplement to your diet is likely to be only a waste of your money.

S-ADENOSYLMETHIONINE (SAMe)

S-adenosylmethionine (SAMe) was first described in 1951 by Giulio Cantoni[191], who recognized it as the primary donor of methyl groups in all living organisms. It is an ancient substance. SAMe and the chain of enzymes necessary to synthesisze it are found in organisms as primitive as *Escherichia coli* bacteria. These enzymes are well-conserved through evolution, with the Methionine adenosyltransferase enzymes of humans and *E.coli* sharing 59 percent of their amino acid sequences[192]. SAMe is found in all cells of the body, and it is involved in dozens of critical biochemical interactions in which single carbons are transferred from one molecule to another. It is a major component of the methylation reactions mediated by folate and vitamin B12, and it plays a role in methylations of protein, DNA, phospholipids, and neurotransmitters. It is SAMe that supplies the methyl groups that convert norepinephrine to epinephrine, and that deactivate catecholamines through the catecholamine-O-methyl transferase reaction. It is also the methyl donor in the synthesis of the choline component of acetylcholine, and in the O-methylation of N-acetyl 5-hydroxytryptamine that converts it to melatonin. Inborn errors and acquired deficiencies in SAMe activity can result in serious pathology. It has recently been appreciated that deficiency of SAMe, perhaps secondary to lack of sufficient B12 and folate, may contribute to the development of Alzheimer's Disease. SAMe has been found to have both antidepressant properties and ability to reverse some of the adverse effects of Metabolic Syndrome on liver function. Thus, SAMe may be a link between Metabolic Syndrome, affective illnesses, and degenerative neurological disorders.

SAMe is synthesized through the action of the enzyme methionine adenosyl transferase that adds ATP to methionine. SAMe then enters the transmethylation cycle, which occurs in various forms in tissues throughout the body. The basic transmethylation process involves the enzymatic transfer of a methyl group from SAMe to a target molecule, with the result being methylation of the target and generation of s-adenosylhomocysteine. The s-adenosylhomocysteine is then hydrolyzed into adenosine and homocysteine. Methionine can be regenerated when N-methyltetrahydrofolate, a form of folic acid, donates its methyl group to homocysteine with help from vitamin B12 and one of several types of methyltransferase enzymes. When folate or B12 is deficient, this transformation of homocysteine back into methionine cannot occur. The body may then suffer both from a lack of methionine for regeneration of SAMe, as well as from elevated serum levels of homocysteine. Hyperhomocysteinemia is a common feature of Metabolic Syndrome[193], and it is considered to be a risk factor for cardiovascular disease. Homocysteine levels also tend to be high in patients with schizophrenia, and it has been suggested that homocysteine may exacerbate the illness[194]. Hyperhomocysteinemia is probably best addressed by replenishing folate and B12, rather than adding SAMe.

Studies from as early as 1975 have shown that SAMe can be as effective as standard antidepressants, including tricyclics, in treating depression[195]. Since that time, a number of studies have shown SAMe to be effective as monotherapy or as augmentation therapy in patients showing minimal response to standard antidepressants[196]. Cerebrospinal fluid levels of SAMe are low in patients with depression[197], and successful treatment of depression is associated with increases in serum levels of this substance[198]. There has also been an interesting report that a particular polymorphism of methylenetetrahydrofolate reductase, the enzyme that activates the methyl group of tertrahydrofolate that is later used in resynthesizing s-adenosylmethionine, is overrepresented in patients with late-onset forms of Major Depression[199]. Thus, there is evidence that deficiency of SAMe may cause or exacerbate depression.

It is not entirely clear how SAMe produces its antidepressant effect. In view of the relationship between SAMe, folate and B12, as well as the fact that folate and B12 deficiencies are relatively common in Major Depression, it is possible that SAMe indirectly treats depression by relieving symptoms that are metabolically downstream from folate and B12 deficiency. It has been suggested that SAMe might improve mood by enhancing dopamine synthesis. However, the mechanism by which this could occur is not obvious. SAMe mediated methylation of dopamine receptors and O-methylation of catecholamines would tend to diminish dopamine effects in the brain. In fact, SAMe can produce Parkinsonian symptoms in experimental animals[200]. SAMe is also said to increase serotonin synthesis in the brain[201]. It is possible that SAMe affects both catecholamine and serotonin synthesis indirectly by enhancing the activity of tetrahydrobiopterin[202], which is involved in several important hydroxylation steps in neurotransmitter synthesis[203]. I do not find this possibility to be terribly compelling.

An interesting possibility is that the antidepressant effect of SAMe could be due in part to SAMe-mediated increases in the synthesis of polyamines in the brain. The polyamines, with charming names such as putrescine and spermine, are synthesized in the brain from SAMe through the action of the enzyme s-adenosylmethinone decarboxylase[204]. Concentrations of the polyamines in the brain decrease in animal models of depression, and are restored by therapeutic doses of SAMe[205]. Intraperitoneal and intracerebroventricular administrations of putrescine have antidepressant effects in the forced swimming and tail-suspension models of depression, and this is thought to be mediated by polyamine activity at central NMDA receptors[206]. It seems possible that SAMe produces antidepressant effects by several different mechanisms.

Also of interest are recent studies showing that SAMe levels are low in cerebrospinal fluid and brain tissue of patients with Alzheimer's Disease[207]. In such patients, SAMe may be reduced by up to 85 percent in some areas of the brain. Activity of the enzyme s-adenosylmethionine decarboxylase, which initiates the transformation of the methionine in SAMe into polyamines, is substantially increased in the brains of Alzheimer's patients[208]. This increase in enzyme activity might further reduce SAMe levels. Sufferers of Alzheimer's are not an overly cheerful population, thus the increase in polyamines does not seem to offer them much respite from depression.

SAMe is quite active in the brain, where it is involved in synthesis and modulation of several important neurotransmitters. Because SAMe contributes methyl groups for the synthesis of choline[209], deficiency of SAMe might contribute to the diminished level of acetylcholine that characterizes Alzheimer's Disease. SAMe also plays an important role in the control of gene expression in the brain through participating in the methylation of DNA. SAMe methylates and thus limits the activities of two important genes responsible for controlling expression of the amyloid protein that forms plaques in Alzheimer's Disease. It has been hypothesized that decreases in SAMe-mediated methylation of the presenilin 1 and beta-secretase genes, may allow overexpression of "sticky" forms of amyloid and progression of Alzheimer's Disease[210]. Other data suggest that increases in s-adenosylhomocysteine, which occurs in the absence of sufficient folate, B12 and SAMe levels, inhibits methylation of the enzyme, protein phosphatase 2A, which in turn disinhibits proliferation of the hyperphosphorylated tau that forms the neurofibrillary tangles of Alzheimer's Disease[211]. In human cell cultures, addition of SAMe has been found to reduce amyloid production[212]. Unfortunately, the only published trial of using SAMe to treat active Alzheimer's Disease showed a lack of efficacy[213]. It is tempting to speculate that earlier intervention may have been more successful. Nonetheless, it is important to note that not all studies have shown decreases in SAMe or its metabolites of brain tissue from Alzheimer's patients[214]. Further studies need to be performed to clarify what, if any, role SAMe might play in the etiology of Alzheimer's Disease.

Folate and B12 play fundamental roles in transmethylation and methionine metabolism. In patients who are deficient, supplementation with folate and B12 can improve insulin sensitivity and decrease the hyperhomocysteinemia that raises cardiovascular risk in Metabolic Syndrome[215]. However, aside from being a mediator of the effects of folate and B12, there is no compelling evidence that SAMe itself has a role in treating Metabolic Syndrome. In a vast literature, there is only one report of SAMe improving insulin sensitivity in an animal model of diabetes type II[216]. In fact, there is evidence that SAMe could aggravate hypertension in Metabolic Syndrome by contributing a methyl group in the synthesis of the natural nitric oxide inhibitor, dimethylarginine[217]. There are data showing that SAMe levels are reduced in the erythrocytes of insulin-resistant patients with diabetes type II. However, the primary problem of methionine/homocysteine metabolism in such patients is a dimished capacity to clear homocysteine from the body, which, ostensibly, contributes to hyperhomocysteinemia. In normal subjects, hyperinsulinemia increases clearance of homocysteine, whereas in patients with diabetes type II, this effect is blunted[218].

The most obvious use of SAMe in patients with Metabolic Syndrome is in cases where the condition has progressed to the point that liver function has been compromised due to Non-alcoholic Fatty Liver Disease (NAFLD) or Non-alcoholic Steatohepatitis (NASH). The liver plays a major role in the development and progression of Metabolic Syndrome. Insulin resistance, stress, and sympathetic hyperactivation push the liver to pump out pathologically large amounts of glucose, VLDL and triglyceride. The liver itself soon begins to labor under the task of processing the ever increasing amounts of fat flooding into it through the portal system. Eventually, it succumbs to the adverse conditions of Metabolic Syndrome and loses its ability to safely handle fatty acids. Fat begins to accumulate in liver tissue where it causes both lipotoxic damage and mechanical damage due to interference of the flow of blood through the delicate sinusoid systems. This is the initiation of NAFLD. It has been estimated that 20 percent of adults in the United States have some degree of NAFLD[219]. If allowed to progress, NAFLD evolves into steatohepatitis (NASH). Inflammation and loss of function is what defines NASH and distinguishes it from the mere accumulation of fatty acid and elevated transaminases in NAFLD. However, the distinction between NAFLD and NASH is somewhat arbitrary.

SAMe is necessary for normal liver function, and essential to prevent liver damage under a variety of pathological conditions. The normal human liver produces about 5 g of SAMe a day[220]. Mice that are genetically engineered to not express MAT type I, the major enzyme involved in synthesizing SAMe from methionine and ATP, spontaneously develop NASH[221]. Their livers show signs of significant oxidative and lipotoxic damage. SAMe helps prevent the production and adverse effects of inflammatory cytokines in liver tissue, and is known to inhibit TNF-alpha production by leukocytes[222] and Kuppfer cells[223], which are resident immune cells in the liver. TNF-alpha may play a role in NASH, Metabolic Syndrome, and some symptoms of Major Depression.

SAMe also protects the liver by serving as substrate for synthesis of the antioxidant glutathione. Although legitimate concern is given to hyperhomocysteinemia, it should be recognized that s-adenosylmethionine participates in the synthesis of glutathione by way of demethylation to s-adenosylhomocysteine. Both a surplus and a lack of homocysteine pose serious problems for the health of the liver. However, it is when folate and B12 deficiency prevents transmethylation and conversion of residual homocysteine back into methionine that hyperhomocysteinemia becomes a problem.

Supplementation with SAMe may help prevent progression of NAFLD to NASH, and it has been suggested that SAMe might be useful in patients showing evidence of having developed NAFLD[224]. However, addressing the underlying causes of the NAFLD, that is, insulin resistance, stress, poor diet, and so on, may be the most prudent first step. When there is evidence of liver damage and loss of function, the use of SAMe as a supplement becomes a far more reasonable measure. Supplemental SAMe would not only help prevent further deterioration of liver function, but also maintain serum levels of SAMe that the liver may no longer be able to generate. If the patient is suffering Major Depression along with NAFLD/NASH, then supplementing with SAMe will not only protect the liver, but might, due to its antidepressant effects, also help resolve an otherwise treatment-resistant psychiatric condition.

Maintaining adequate folate and B12 levels will generally obviate any need for supplementation with SAMe. I have heard it stated quite emphatically that folate and B12 supplementation should be sufficient treatment under any circumstances, and that use of SAMe is simply a waste of money. However, it is important to realize that the transmethylation processes that mediate the beneficial effects of folate and vitamin B12 are driven by enzymes and require intact tissue. The liver is the site where nearly 85 percent of the transmethylation involving SAMe takes place, and these processes can be severely compromised in patients with liver disease. Thus, an important use for SAMe in individuals with both Metabolic Syndrome and Major Depression may be in helping maintain normal liver function in individuals whose livers have succumbed to the hepatic component of Metabolic Syndrome, that is, non-alcoholic fatty liver disease (NAFLD).

Although I have no proof of efficacy, I have several times added SAMe to the antidepressant treatment of patients receiving interferon for treatment of hepatitis C. In some cases, patients who had previously become profoundly depressed and thus unable to tolerate interferon treatment despite being on an antidepressant, were able to complete that treatment after SAMe was added. In view of the evidence that SAMe can benefit patients with liver disease and help alleviate depression with few if any adverse effects, I would think it worthwhile to have some controlled studies of the use of SAMe as an augmentation strategy for prophylaxis against depression in patients being treated for hepatitis C with interferon.

SAMe is generally given in doses of 400–1600 mg a day. Doses as high as 1600 mg a day have been found to be well tolerated and essentially free of toxic effects[225]. The only concerns to bear in mind are that supplying large amounts of SAMe in the absence of adequate folate and B12 would tend to increase serum homocysteine. High doses of SAMe might also aggravate Parkinsonian symptoms in susceptible individuals, or, like other antidepressants, induce mania in patients with BPAD that are predisposed to "switching". Also, keep in mind that SAMe is quite expensive and almost never covered by insurance. If serum folate and B12 levels are adequate in patients with normal liver function, then SAMe may be unnecessary. Still, there is reason to believe that SAMe may act independently of folate and B12 by serving as raw material to produce polyamines in the brain. Moreover, if there is evidence of NAFLD, hepatitis, interferon treatment, or other conditions in which the enzymatic processes of synthesizing SAMe in liver are compromised, then addition of SAMe may be useful in achieving relief from symptoms in patients who might otherwise remain resistant to treatment. Finally, it is worth noting that betaine[226], another natural methyl donor, may offer benefits similar to those of SAMe. Betaine is far less expensive than SAMe and likely more bioavailable.

SILYMARIN

Silymarin is a substance extracted from the milk thistle, that has been used as a healing substance for the last 2000 years[227]. It has primarily been used for treatment of liver diseases, and it appears to be quite safe even in daily use. Reports in the scientific literature are somewhat mixed in respect to the effectiveness of silymarin in treating patients with cirrhosis of the liver secondary to alcoholism[228]. However, more clearly positive results have come out of studies using silymarin to treat liver damage secondary to poisoning from the lethally toxic mushroom *Amanita phalloides*. What may be equally promising are the effects of silymarin that act to reverse some components of Metabolic Syndrome.

Silymarin appears to reduce inflammation, which is a major component of Metabolic Syndrome. It blocks the activation NFk-B, which is one of the cell factors that turns on the inflammatory response in the body[229]. NFk-B also contributes to insulin resistance[230]. Silymarin also reduces activity of TNF-alpha, which not only stimulates NFk-B, but also causes immune system cells to release reactive oxygen compounds that causes oxidative damage and further deficits in the responses to insulin. In a study of patients diagnosed with both cirrhosis and diabetes, silymarin was found to reduce the abnormal oxidation of fats, increase sensitivity to insulin, and lower abnormally high levels of insulin in the blood[231]. Silymarin can also help protect the pancreas from a number of toxins that damage the beta cells through oxidative injury[232].

Among the primary signs of Metabolic Syndrome are elevations in serum triglycerides and bad cholesterols, in the form of VLDL and LDL cholesterol, and reductions in the good cholesterol, HDL. There are studies that suggest that silymarin can inhibit the synthesis of VLDL and LDL cholesterol[233], lower triglycerides, and increase HDL[234]. The anti-oxidant effects of silymarin allow it to help reduce oxidation of LDL cholesterol[235], the products of which cause build up of blockages in the coronary arteries[236].

The molecular structure of silymarin, referred to by biochemists as "polyphenolic", gives it the property of being able to pass from the blood into the brain. In some animal studies, silymarin was found to spare the brain from damaging effects of oxygen radicals after the blood supply was cut off for 30 minutes[237]. It is reasonable to expect that silymarin may provide a little help in protecting the brain from the inflammatory and oxidative stress effects by which Metabolic Syndrome contributes to Alzheimer's Disease.

Silymarin may be a useful supplement for individuals with signs of Metabolic Syndrome, particularly those with any type of compromise to liver function, such as hepatitis C, alcoholic cirrhosis, or fatty liver. Silymarin, and similar substances found in milk thistle such as silibin, are available commercially. The usual dose is about 50 mg of extract several times a day.

VANADIUM

Vanadium is a metalic element that appears as number 23 in the periodic table. For many years it has been suspected to be essential in human physiology[238]. There have been strong arguments that very small amounts of vanadium are necessary to maintain health. However, the microgram quantities of vanadium that are likely to be essential for human health are far smaller than the milligram doses of vanadium that can be used to improve glucose metabolism and insulin sensitivity. Thus, while vanadium may be an essential trace mineral, it may also be used at high doses as a drug.

Vanadium is found in high concentrations in the blood of primitive sea creatures, such as sea squirts. It makes their blood green. For many years it was suspected that vanadium carries oxygen in their blood, just as iron binds and carries oxygen in the blood of higher animals. However, since vanadium is poor in binding oxygen, and these primitive creatures do have some iron in their blood, the vanadium likely serves some other purpose for these animals. Perhaps the high concentrations of vanadium simply serve as a form of toxin to dissuade more nimble, intelligent creatures from eating them.

For over 100 years it has been known that vanadium can improve symptoms of diabetes. Over the last 20 years, studies have clearly established the fact that oral vanadium can lower serum glucose, reduce levels of fasting insulin, and increase insulin sensitivity in human subjects with diabetes type II[239]. It is still not clear how vanadium acts to improve symptoms of diabetes. Does it mimic the effects of insulin, enhance the effects of insulin already present, or do a little of both? Vanadium has little if any effect on serum insulin or glucose levels in healthy human subjects without diabetes[240]. It appears that vanadium cannot go above and beyond what normal insulin action does.

Like insulin, vanadium lowers serum glucose by both enhancing the uptake of glucose into muscle cells and reducing the output of glucose by the liver. Vanadium further acts like insulin in its ability to stimulate fat storage in adipose cells and to reduce the break down of fat already in storage[241]. These effects reduce serum glucose by forcing muscle cells to rely on glucose rather than fatty acids as fuel. Vanadium, like insulin, may also act in the brain to reduce appetite. The hormone neuropeptide Y is known to increase appetite by acting in an area of the brain called the hypothalamus. In animal studies, both insulin and vanadium have been found to decrease concentrations of neuropeptide Y in the hypothalamus[242].

A study from Japan revealed that concentrations of vanadium equal to those found in ground water from areas of high volcanic activity can enhance perception of sweetness[243]. It is not known how this occurs, or if it has anything to do with vanadium's ability to alter glucose metabolism. However, it would be interesting if vanadium supplements could both improve the symptoms of diabetes and lower the intake of sugars through reducing the amount of sugar required to produce a sweet taste.

Vanadyl sulphate and sodium orthovanadate are effective in reducing glucose levels and increasing insulin sensitivity in some individuals[244]. These simple, inorganic forms of vanadium are not readily absorbed from the intestine, and can cause stomach upset and diarrhea. In some individuals, the doses needed to produce meaningful improvements in glucose metabolism are not tolerated. Unfortunately, the more exotic, organically complexed forms of vanadium that are more readily absorbed and better tolerated are not available outside of research laboratories. There are many individuals who have taken oral vanadyl sulphate at doses of 50 mg twice a day without gastrointestinal symptoms[245].

Not all researchers having glowing reports about the effects of vanadium in human subjects. Up to 300 mg a day of vanadyl sulphate were found to be safe in human diabetic subjects. However, it did not dramatically improve insulin sensitivity or glycemic control[246]. Moreover, whereas studies dating as far back as the 1950s have suggested that vanadium might be useful in lowering cholesterol, this reduction in cholesterol is primarily a lowering of HDL. Vanadium may also increase serum triglycerides[247].

A final caution must be voiced in the context of some old reports of a relationship between serum vanadium levels and Major Depressive Illness. Several reports from 1984 to 1987 suggested that serum levels of vanadium are increased in individuals with severe Bipolar Affective Disorder and other severe forms of depression accompanied by psychosis[248]. In some cases, levels of vanadium were found to decrease as patients improved. Adverse effects of vanadium on the enzyme, sodium potassium ATPase, can be reversed by lithium, a drug with beneficial effects on depression and mania. This might provide a potential mechanism by which vanadium could affect brain activity and, in turn, mood. Nonetheless, virtually nothing more has been reported over the last 20 years to further implicate vanadium in psychiatric illness.

In view of the fact that vanadium has rather mixed effects in treating diabetes type II, I would be extremely reluctant to recommend using the readily available forms of vanadium, that is, vanadyl sulfate and sodium orthovanadate, in the treatment of Metabolic Syndrome. There are too many other nutritional supplements to take, and dietary changes to institute, that are both helpful and without ill effects. The more easily absorbed forms of vanadium may be available in the near future, and wider clinical trials may provide evidence of efficacy at well tolerated doses.

VITAMIN D

Vitamin D is primarily thought of as "the sunshine vitamin" critical in calcium metabolism and necessary for the development and maintenance of strong bones. Even many physicians are unaware of the variety and importance of vitamin D's activities in the body. There is growing evidence that vitamin D may have a role in the prevention and treatment of Metabolic Syndrome, Alzheimer's Disease, Major Depression, Seasonal Affective Disorder and other psychiatric illnesses.

What is referred to as "vitamin D" is actually a family of related substances. The underlying structure of vitamin D is cholesterol. The provitamin, 7-dehydrocholesterol, is synthesized in the skin. The exposure of 7-dehydrocholesterol to ultraviolet light opens the bond between carbons 9 and 10 in one of the four sterol rings and produces cholecalciferol, which is better known as vitamin D3. Vitamin D2 is produced in plants by ultraviolet irradiation of the cholesterol-like substance ergosterol. Vitamin D2 is available to us from consumption of plants or the meat and milk of herbivorous animals. The human body can use either vitamin D2 or D3 as the starting point for the various physiologically active metabolites. After synthesis in the skin or absorption from the gut, vitamin D is taken up by the liver and transformed into 25-hydroxyvitamin D. 25-hydroxyvitamin D is transferred to the kidney where it is transformed into 1,25-dihydroxyvitamin D, which is the most potent form of vitamin D.

The mechanism of action of vitamin D is essentially that of a steroid hormone. It first binds to a receptor in the cytosol of the target cell. It is then transported to the nucleus where it binds to the DNA. Binding to the DNA stimulates the synthesis of calbindin and other proteins that mediate the effects of vitamin D. Some effects of vitamin D are too rapid to be explained by the action at DNA promoter sites, and it has been suspected that the vitamin also acts directly on receptors in cell membranes.

The best known effects of vitamin D are the enhancement of absorption of calcium from the gut and control of the deposition of minerals in the bones. However, in 1980 it was discovered that deficiency of vitamin D in rats inhibits secretion of insulin from the pancreas[249]. This was later found to be true in humans[250]. Deficiency in vitamin D results in poor glucose tolerance, whereas repletion helps reverse the intolerance[251]. Vitamin D deficiency increases the risk of diabetes type II[252]. There are inconsistencies in the literature about the effect of vitamin D on insulin sensitivity[253]. However, several studies have shown that vitamin D deficiency increases insulin resistance[254] and the risk of developing Metabolic Syndrome[255]. Vitamin D deficiency may also contribute to hypertension[256]. However, supplemental vitamin D does not relieve hypertension in people who are not deficient in the vitamin[257].

In rat brain, vitamin D receptors are found throughout the limbic system, with particularly high densities in the hippocampus[258]. Those structures are involved in the processing of emotion and memory. This distribution of vitamin D receptors may explain the effects of vitamin D on mood and cognitive function. Given that Seasonal Affective Disorder (SAD) can be treated with sunlight, and vitamin D is "the sunshine vitamin", it seemed almost too obvious that the vitamin could relieve symptoms of this cyclic mood disorder. However, an impressive collection of studies has shown that vitamin D is useful for the treatment of SAD. In one case, a direct comparison between light therapy and treatment with vitamin D found the vitamin more effective[259]. It is not clear if the large dose of vitamin D used in this study, 100 000 IU, is necessary and safe for general clinical application. Indeed, others have found that more reasonable doses of 400–800 IU of vitamin D are effective for treating SAD[260]. Vitamin D levels are significantly lower in patients with schizophrenia, major depression and alcoholism than in healthy control subjects[261]. It is not clear if these relative deficiencies contribute to those illnesses, or if they are the result of the poor health maintenance that often accompanies them.

There is not yet any definitive evidence that vitamin D deficiency causes or contributes to Alzheimer's Disease. However, there are reasons to suspect that it could be helpful in preventing or treating the disease. Many sufferers of Alzheimer's Disease are deficient in vitamin D[262], and their brains have relatively low numbers of vitamin D receptors[263]. Vitamin D may help protect the brain against the damage caused by chronic, high levels of stress hormones. In rat brain, pretreatment with vitamin D prevented the shrinkage and deaths of hippocampal neurons that occurred with exposure to the cortisol-like drug dexamthethasone[264]. Other studies using rats have found vitamin D to help maintain the production of acetylcholine in specific areas of the brain[265]. Both of the above effects could be expected to help protect against Alzheimer's Disease if vitamin D has the same effects in human brain.

The human body is perfectly capable of synthesizing 100 percent of the vitamin D it needs by exposure of the skin to sunlight[266]. However, many people do not get sufficient exposure to sunlight. Working indoors and wearing business dress that reveals little skin can be part of the problem. Of course, exposure to sunlight varies considerably as the seasons change. People expose less skin when covering up against exposure to the cold of winter. Decrease in day length during the winter, as well as the sun tracking low in the sky in extreme latitudes also dramatically reduces intensity of sunlight and the rate of synthesis of vitamin D. Although spring and summer are the seasons of sunlight, legitimate concern about melanoma has led to widespread use of sunscreen lotions. Most sunscreens can quite effectively block the UVB that converts 7-dehydrocholesterol to D3. It has been reported that even a weak SPF 8 lotion can almost completely eliminate the synthesis of vitamin D in the skin[267].

Obtaining vitamin D from the diet can obviate the problem of low exposure to sunlight. Unfortunately, many adults do not consume sufficient amounts of vitamin D-containing food products, such as fortified milk. Consequently, vitamin D deficiency is far more prevalent than many physicians suspect. In a study of healthy, young hospital workers in Boston, 36 percent of them were deficient in vitamin D by the end of the winter[268]. A recent review reveals that as many as 57 percent of general medicine inpatients in the United States suffer from inadequate levels of vitamin D[269]. In the elderly and nursing home patients, the percentage may be higher.

The solution to vitamin D deficiency is simple. A healthy individual need only take 10–20 μg, or 400–800 IU of vitamin D daily to prevent deficiency.

Because vitamin D can be toxic in high quantity, there is no compelling reason to take more than 800 IU a day. People with severe deficiencies, or diagnoses of liver or kidney ailments need medical evaluation and supervision for adequate and safe vitamin D replacement.

REFERENCES

1. Rebouche, C.J., Carnitine. In *Nutrition in Health and Disease* (9th ed.), M.E. Shils, J.A. Olson, M. Shike and A.C. Ross, eds. Williams & Wilkins, Baltimore, 1999, pp. 505–512.
2. Costell, M., O'Connor, J.E., and Grisolia, S., Age-dependent decrease of carnitine content in muscle of mice and humans. *Biochem. Biophys. Res. Commun.* 1989; 161:1135–1143.
3. Montgomery, S.A., Thal, L.J., and Amrein, R., Meta-analysis of double blind randomized controlled clinical trials of acetyl-L-carnitine versus placebo in the treatment of mild cognitive impairment and mild Alzheimer's Disease. *Int. Clin. Psychopharmacol.* March 2003; 18(2):61–71.
4. Pettegrew, J.W. et al., 31P-MRS study of acetyl-L-carnitine treatment in geriatric depression: preliminary results. *Bipolar Disord.* 2002; 4:61–66.
5. Vermeulen, R.C. and Scholte, H.R., Exploratory open label, randomized study of acetyl- and propionylcarnitine in chronic fatigue syndrome. *Psychosom. Med.* 2004; 66:276–282.
6. Capaldo, B. et al., Carnitine improves peripheral glucose disposal in non-insulin-dependent diabetic patients. *Diabetes Res. Clin. Pract.* 1991; 14:191–195.
7. Sima, A.A. et al., Acetyl-L-carnitine improves pain, nerve regeneration, and vibratory perception in patients with chronic diabetic neuropathy: an analysis of two randomized placebo-controlled trials. *Diabetes Care* 2005; 28:89–94.
8. Wada, R. and Yagihashi, S., Role of advanced glycation end products and their receptors in development of diabetic neuropathy. *Ann. N.Y. Acad. Sci.* 2005; 1043:598–604.
9. Abdul, H.M. et al., Acetyl-L-carnitine-induced up-regulation of heat shock proteins protects cortical neurons against amyloid-beta peptide 1-42-mediated oxidative stress and neurotoxicity: Implications for Alzheimer's Disease. *J. Neurosci. Res.* 2006; 84:398–408.
10. Tanaka, Y. et al., Acetyl-L-carnitine supplementation restores decreased tissue carnitine levels and impaired lipid metabolism in aged rats. *J. Lipid. Res.* 2004; 45:729–735.
11. Albertazzi, A. et al., Endocrine-metabolic effects of l-carnitine in patients on regular dialysis treatment. *Proc. Eur. Dial. Transplant. Assoc.* 1983; 19:302–307.
12. Jacob, S. et al., Oral administration of RAC-alpha-lipoic acid modulates insulin sensitivity in patients with type-2 diabetes mellitus: a placebo-controlled pilot trial. *Free Radic. Biol. Med.* 1999; 27:309–314.
13. Konrad, D., Utilization of the insulin-signaling network in the metabolic actions of alpha-lipoic acid-reduction or oxidation? *Antioxid. Redox Signal.* 2005; 7:1032–1039.
14. Konrad, D. et al., The antihyperglycemic drug alpha-lipoic acid stimulates glucose uptake via both GLUT4 translocation and GLUT4 activation: potential role of p38 mitogen-activated protein kinase in GLUT4 activation. *Diabetes* 2001; 50:1464–71.
15. Lee, W.J. et al., Alpha-lipoic acid increases insulin sensitivity by activating AMPK in skeletal muscle. *Biochem. Biophys. Res. Commun.* 2005; 332:885–891.
16. Ford, I. et al., The effects of treatment with alpha-lipoic acid or evening primrose oil on vascular hemostatic and lipid risk factors, blood flow, and peripheral nerve conduction in the streptozotocin-diabetic rat. *Metabolism* 2001; 50:868–875.
17. Zhang, W.J. and Frei, B., Alpha-lipoic acid inhibits TNF-alpha-induced NF-kappaB activation and adhesion molecule expression in human aortic endothelial cells. *FASEB J.* 2001; 15:2423–2432.
18. Hipkiss, A.R., Glycation, ageing and carnosine: Are carnivorous diets beneficial? *Mech. Ageing Dev.* 2005; 126:1034–1039.
19. Gayova, E., Carnosine in patients with type I diabetes mellitus. *Bratisl. Lek. Listy.* 1999; 100:500–502.
20. Krajcovicova-Kudlackova, M. et al., Advanced glycation end products and nutrition. *Physiol. Res.* 2002; 51:313–316.

21. Klebanov, G.I. et al., Effect of carnosine and its components on free-radical reactions. *Membr. Cell. Biol.* 1998; 12:89–99.
22. Guilaeva, N.V. et al., Carnosine prevents the activation of free-radical lipid oxidation during stress. *Biull. Eksp. Biol. Med.* 1989; 107:144–147.
23. Niijima, A. et al., Effects of L-carnosine on renal sympathetic nerve activity and DOCA-salt hypertension in rats. *Auton. Neurosci.* 2002; 97:99–102.
24. Jackson, M.C. and Lenney, J.F., The distribution of carnosine and related dipeptides in rat and human tissues. *Inflamm. Res.* 1996; 45:132–135.
25. Preston, J.E. et al., Toxic effects of beta-amyloid(25-35) on immortalised rat brain endothelial cell: protection by carnosine, homocarnosine and beta-alanine. *Neurosci. Lett.* 1998; 242:105–108.
26. Hipkiss, A.R., Is carnosine a naturally occuring supressor of oxidative damage in olfactory neurons? *Rejuvenation Res.* 2004; 7:253–255.
27. Harris, R.C. et al., The absorption of orally supplied beta-alanine and its effect on muscle carnosine synthesis in human vastus lateralis. *Amino Acids* 2006; 30:279–289.
28. Park, Y.J. et al., Quantitation of carnosine in human plasma after dietary consumption of beef. *J. Agric. Food Chem.* 2005; 15:4736–4739.
29. Chan, K.M. and Decker, E.A., Endogenous skeletal muscle antioxidants. *Crit. Rev. Food Sci. Nutr.* 1994; 34:403–426.
30. Marchis, S.D. et al., Carnosine-related dipeptides in neurones and glia. *Biochemistry (Mosc.)* 2000; 65:824–833.
31. Ding, E.L. et al., Chocolate and prevention of cardiovascular disease: a systematic review. *Nutr. Metab. (Lond.)* 2006; 3:2.
32. Grassi, D. et al., Short-term administration of dark chocolate is followed by a significant increase in insulin sensitivity and a decrease in blood pressure in healthy persons. *Am. J. Clin. Nutr.* 2005; 81:611–614.
33. Karim, M., McCormick, K., and Kappagoda, C.T., Effects of cocoa extracts on endothelium-dependent relaxation. *J. Nutr.* 2000; 130(8S Suppl.):2105S–2108S.
34. Wegener, G. et al., Local, but not systemic, administration of serotonergic antidepressants decreases hippocampal nitric oxide synthase activity. *Brain Res.* 2003; 959:128–134.
35. Schramm, D.D. et al., Chocolate procyanidins decrease the leukotriene-prostacyclin ratio in humans and human aortic endothelial cells. *Am. J. Clin. Nutr.* 2001; 73:36–40.
36. Ziegleder, G., Stojacic, E., and Stumpf, B., [Occurrence of beta-phenylethylamine and its derivatives in cocoa and cocoa products] *Z Lebensm Unters Forsch* 1992; 195:235–238.
37. Sabelli, H. et al., Phenylethylamine relieves depression after selective MAO-B inhibition. *J. Neuropsychiatry Clin. Neurosci.* Spring 1994; 6:203.
38. Wainscott, D.B. et al., Pharmacologic characterization of the cloned human trace amine-associated receptor1 (TAAR1) and evidence for species differences with the rat TAAR1. *J. Pharmacol. Exp. Ther.* 2007; 320:475–485.
39. James, J.S., Marijuana and chocolate. *AIDS Treat News* 1996; 257:3–4.
40. Michener, W. and Rozin, P., Pharmacological versus sensory factors in the satiation of chocolate craving. *Physiol. Behav.* 1994; 56:419–422.
41. Kris-Etherton, P.M. and Mustad, V.A., Chocolate feeding studies: a novel approach for valuating the plasma lipid effects of stearic acid. *Am. J. Clin. Nutr.* 1994; 60(Suppl. 6) 1029S–1036S.
42. Mertz, W., Chromium research from a distance: from 1959 to 1980. *J. Am. Coll. Nutr.* 1998; 17:544–547.
43. Vincent, J.B., Recent advances in the nutritional biochemistry of trivalent chromium. *Proc. Nutr. Soc.* 2004; 63:41–47.
44. Vincent, J.B., The biochemistry of chromium. *J. Nutr.* 2000; 130:715–718.
45. Anderson, R.A., Chromium in the prevention and control of diabetes. *Diabetes Metab.* 2000; 26:22–27.
46. Lee, N.A. and Reasner, C.A., Beneficial effect of chromium supplementation on serum triglyceride levels in NIDDM. *Diabetes Care* 1994; 17:1449–1452.
47. Bahijiri, S.M. et al., The effects of inorganic chromium and brewer's yeast supplementation on glucose tolerance, serum lipids and drug dosage in individuals with type 2 diabetes. *Saudi Med. J.* 2000; 21:831–837.

48. Althius, M.D. et al., Glucose and insulin responses to dietary chromium supplements: a meta-analysis. *Am. J. Clin Nutr.* 2002; 76:148–155.

49. Clodfelder, B.J., Upchurch, R.G., and Vincent, J.B., A comparison of the insulin-sensitive transport of chromium in healthy and model diabetic rats. *J. Inorg. Biochem.* 2004; 98:522–533.

50. Hepburn, D.D. et al., Nutritional supplement chromium picolinate causes sterility and lethal mutations in Drosophila melanogaster. *Proc. Natl. Acad. Sci. U.S.A.* 2003; 100:3766–3771.

51. Stearns, D.M. et al., Chromium(III) picolinate produces chromosome damage in Chinese hamster ovary cells. *FASEB J.* 1995; 9:1643–1648.

52. Kato, I. et al., Effect of supplementation with chromium picolinate on antibody titers to 5-hydroxymethyl uracil. *Eur. J. Epidemiol.* 1998; 14:621–626.

53. Folkers, K. et al., Lovastatin decreases coenzyme Q levels in humans. *Proc. Natl. Acad. Sci. U.S.A.* 1990; 87:8931–8934.

54. Singh, R.B. et al., Effect of hydrosoluble coenzyme Q10 on blood pressures and insulin resistance in hypertensive patients with coronary artery disease. *J. Hum. Hypertens.* 1999; 13:203–208.

55. Miles, M.V. et al., Coenzyme Q10 changes are associated with metabolic syndrome. *Clin. Chim. Acta* 2004; 344:173–179.

56. Beal, M.F., Mitochondrial dysfunction and oxidative damage in Alzheimer's and Parkinson's diseases and coenzyme Q10 as a potential treatment. *J. Bioenerg. Biomembr.* 2004; 36:381–386.

57. Pari, L. and Murugan, P., Effect of tetrahydrocurcumin on blood glucose, plasma insulin and hepatic key enzymes in streptozotocin induced diabetic rats. *J. Basic Clin. Physiol. Pharmacol.* 2005; 16:257–274.

58. Babu, P.S. and Srinivasan, K., Hypolipidemic action of curcumin, the active principle of turmeric (Curcuma longa) in streptozotocin induced diabetic rats. *Mol. Cell. Biochem.* 1997; 166:169–175.

59. Nishiyama, T. et al., Curcuminoids and sesquiterpenoids in turmeric (Curcuma longa L.) suppress an increase in blood glucose level in type 2 diabetic KK-Ay mice. *J. Agric. Food Chem.* 2005; 53:959–963.

60. Balasubramanyam, M. et al., Curcumin-induced inhibition of cellular reactive oxygen species generation: novel therapeutic implications. *J. Biosci.* 2003; 28:715–721.

61. Bengmark, S., Curcumin, an atoxic antioxidant and natural NFκB, cyclooxygenase-2, lipooxygenase, and inducible nitric oxide synthase inhibitor: a shield against acute and chronic diseases. *JPEN J. Parenter. Enteral. Nutr.* 2006; 30:45–51.

62. Yang, F. et al., Curcumin inhibits formation of amyloid beta oligomers and fibrils, binds plaques, and reduces amyloid in vivo. *J. Biol. Chem.* 2005; 280:5892–5901.

63. Ringman, J.M. et al., A potential role of the curry spice curcumin in Alzheimer's Disease. *Curr. Alzheimer Res.* 2005; 2:131–136.

64. Chainani-Wu, N., Safety and anti-inflammatory activity of curcumin: a component of tumeric (Curcuma longa). *J. Altern. Complement Med.* 2003; 9:161–168.

65. Gupta, B. et al., Mechanisms of curcumin induced gastric ulcer in rats. *Indian J. Med. Res.* 1980; 71:806–814.

66. Feldman, H.A. et al., Low dehydroepiandrosterone and ischemic heart disease in middle-aged men: prospective results from the Massachusetts Male Aging Study. *Am. J. Epidemiol.* 2001; 153:79–89.

67. Kawano, H. et al., Dehydroepiandrosterone supplementation improves endothelial function and insulin sensitivity in men. *J. Clin. Endocrinol. Metab.* 2003; 88:3190–3195.

68. Villareal, D.T. and Holloszy, J.O., Effect of DHEA on abdominal fat and insulin action in elderly women and men: a randomized controlled trial. *JAMA* 2004; 292:2243–2248.

69. Kalimi, M. et al., Anti-glucocorticoid effects of dehydroepiandrosterone (DHEA). *Mol. Cell. Biochem.* 1994; 131:99–104.

70. Peters, J.M. et al., Peroxisome proliferator-activated receptor alpha required for gene induction by dehydroepiandrosterone-3 beta-sulfate. *Mol. Pharmacol.* 1996; 50:67–74.

71. Karishma, K.K. and Herbert, J., Dehydroepiandrosterone (DHEA) stimulates neurogenisis in the hippocampus of the rat, promotes survival of newly formed neurons and prevents corticosterone-induced suppression. *Eur. J. Neurosci.* 2002; 16:445–453.

72. Wolkowitz, O.M.N. et al., Double-blind treatment of major depression with dehy-droepiandrosterone. *Am. J. Psychiatry* 1999; 156:646–649.

73. Hackbert, L. and Heiman, J.R., Acute dehydroepiandrosterone (DHEA) effects on sexual arousal in postmenopausal women. *J. Womens Health Gend. Based Med.* 2002; 11:155–162.

74. Rhodes, M.E. et al., Enhanced plasma DHEAS, brain acetylcholine and memory mediated by steroid sulfatase inhibition. *Brain Res.* 1997; 773:28–32.

75. Dean, C.E., Prasterone (DHEA) and mania. *Ann. Pharmacother.* 2000; 34:1419–1422.

76. Pitts, R.L., Serum elevation of dehydroepiandrosterone sulfate associated with male pattern baldness in young men. *J. Am. Acad. Dermatol.* 1987; 16(3 Pt 1):571–573.

77. Le, H. et al., Dihydrotestosterone and testosterone, but not DHEA or estradiol, differentially modulate IGF-I, IGFBP - 2 and IGFBP-3 gene and protein expression in primary cultures of human prostatic stromal cells. *Am. J. Physiol. Endocrinol. Metab.* 2006; 290:952–960.

78. Kaaks, R. et al., Postmenopausal serum androgens, oestrogens and breast cancer risk: the European prospective investigation into cancer and nutrition. *Endocr. Relat. Cancer* 2005; 12:1071–1082.

79. Feskens, E.J., Bowles, C.H., and Kromhout, D., Inverse association between fish intake and risk of glucose intolerance in normoglycemic elderly men and women. *Diabetes Care* 1991; 14:935–941.

80. Feskens, E.J. et al., Dietary factors determining diabetes and impaired glucose tolerance. A 20-year follow-up of the Finnish and Dutch cohorts of the Seven Countries Study. *Diabetes Care* 1995; 18:1104–1112.

81. Bulliyya, G., Fish intake and blood lipids in fish eating versus non-fish eating communities of coastal south India. *Clin. Nutr.* 2000; 19:165–170.

82. Oomen, C.M. et al., Fish consumption and coronary heart disease mortality in Finland, Italy, and the Netherlands. *Am. J. Epidemiol.* 2000; 151:999–1006.

83. Weintraub, M.S. et al., Dietary polyunsaturated fats of the W-6 and W-3 series reduce postprandial lipoprotein levels. Chronic and acute effects of fat saturation on postprandial lipoprotein metabolism. *J. Clin. Invest.* 1988; 82:1884–1893.

84. Sanders, T.A., Dietary fat and postprandial lipids. *Curr. Atheroscler. Rep.* 2003; 5:445–451.

85. Griffin, B.A., The effect of n-3 fatty acids on low density lipoprotein subfractions. *Lipids* 2001; 36 (Suppl.):S91–S97.

86. Mori, T.A., Omega-3 fatty acids and hypertension in humans. *Clin. Exp. Pharmacol. Physiol.* 2006; 33:842–846.

87. Delarue, J. et al., N-3 long chain polyunsaturated fatty acids: a nutritional tool to prevent insulin resistance associated to type 2 diabetes and obesity? *Reprod. Nutr. Dev.* 2004; 44:289–299.

88. Giugliano, D., Ceriello, A., and Esposito, K., The effects of diet on inflammation: emphasis on the metabolic syndrome. *J. Am. Coll. Cardiol.* 2006; 48:677–685.

89. Das, U.N., Biological significance of essential fatty acids. *J. Assoc. Physicians India* 2006; 54:309–319.

90. Ferrucci, L. et al., Relationship of plasma polyunsaturated fatty acids to circulating inflammatory markers. *J. Clin. Endocrinol. Metab.* 2006; 91:439–446.

91. Mishra, A., Chaudhary, A., and Sethi, S., Oxidized omega-3 fatty acids inhibit NF-kappaB activation via a PPARalpha-dependent pathway. *Arterioscler. Thromb. Vasc. Biol.* 2004; 24:1621–1627.

92. Carpentier, Y.A., Portois, L., and Malaisse, W.J., n-3 fatty acids and the metabolic syndrome. *Am. J. Clin. Nutr.* 2006; 83(6 Suppl.):1499S–1504S.

93. Uauy, R. and Dangour, A.D., Nutrition in brain development and aging: role of essential fatty acids. *Nutr. Rev.* 2006; 64(5 Pt 2):S24–S33;

94. Glomset, J.A., Role of docosahexaenoic acid in neuronal plasma membranes. *Sci. STKE* 2006; 2006(321):pe6.

95. Delion, S. et al., Chronic dietary alpha-linolenic acid deficiency alters dopaminergic and serotoninergic neurotransmission in rats. *J. Nutr.* 1994; 124:2466–2476.

96. McNamara, R.K. et al., Modulation of phosphoinositide-protein kinase C signal transduction by omega-3 fatty acids: implications for the pathophysiology and treatment of recurrent neuropsychiatric illness. *Prostaglandins Leukot. Essent. Fatty Acids* 2006; 75:237–257.

97. Corporeau, C. et al., Adipose tissue compensates for defect of phosphatidylinositol 3′-kinase induced in liver and muscle by dietary fish oil in fed rats. *Am. J. Physiol. Endocrinol. Metab.* 2006; 290:E78–E86.

98. Tankskanen, A. et al., Fish consumption and depressive symptoms in the general population of Finland. *Psychiatr. Serv.* 2001; 52:529–531.

99. Timonen, M. et al., Fish consumption and depression: the Northern Finland 1966 birth cohort study. *J. Affect. Disord.* 2004; 82:447–452.

100. Jacka, F.N., Dietary omega-3 fatty acids and depression in a community sample. *Nutr. Neurosci.* 2004; 7:101–106.

101. Peet, M., Murphy, B., Shay, J., and Horrobin, D., Depletion of omega-3 fatty acid levels in red blood cell membranes of depressive patients. *Biol. Psychiatry* 1998; 43:315–319.

102. Edwards, R. et al., Omega-3 polyunsaturated fatty acid levels in the diet and in red blood cell membranes of depressed patients. *J. Affect. Disord.* 1998; 48:149–155.

103. Mamalakis, G., Tornaritis, M., and Kafatos, A., Depression and adipose essential fatty acids. *Prostaglandins Leukot. Essent. Fatty Acids* 2002; 67:311–318.

104. Su, K.-P. et al., Omega-3 fatty acids in major depressive disorder – a preliminary double-blind, placebo-controlled trial. *Euro. Neuropsychopharmacol.* 2003; 13:267–271.

105. Silvers, K.M. et al., Randomised double-blind placebo-controlled trial of fish oil in the treatment of depression. *Prostaglandins Leukot. Essent. Fatty Acids* 2005; 72:211–218.

106. Marangell, L.B. et al., A double-blind, placebo-controlled study of the omega-3 fatty acid docosahexaenoic acid in the treatment of major depression. *Am. J. Psychiatry* 2003; 160:996–998.

107. Nemets, B., Stahl, Z., and Belmaker, R.H., Addition of omega-3 fatty acid to maintenance medication treatment for recurrent unipolar depressive disorder. *Am. J. Psychiatry* 2002; 159:477–479.

108. Freeman, M.P., Omega-3 fatty acids and perinatal depression: a review of the literature and recommendations for future research. *Prostaglandins Leukot. Essent. Fatty Acids* 2006; 75:291–297.

109. Appleton, K.M. et al., Effects of n-3 long-chain polyunsaturated fatty acids on depressed mood: systematic review of published trials. *Am. J. Clin. Nutr.* 2006; 84:1308–1316.

110. Noaghiul, S., and Hibbeln, J.R., Cross-national comparisons of seafood consumption and rates of bipolar disorders. *Am. J. Psychiatry* 2003; 160:2222–2227.

111. Chiu, C.C. et al., Polyunsaturated fatty acid deficit in patients with bipolar mania. *Eur. Neuropsychopharmacol.* 2003; 13:99–103.

112. Frangou, S., Lewis, M., and McCrone, P., Efficacy of ethyl-eicosapentaenoic acid in bipolar depression: randomised double-blind placebo-controlled study. *Br. J. Psychiatry* 2006; 188:46–50.

113. Keck, P.E. Jr. et al., Double-blind, randomized, placebo-controlled trials of ethyl-eicosapentanoate in the treatment of bipolar depression and rapid cycling bipolar disorder. *Biol. Psychiatry* 2006; 60:1020–1022.

114. Sagduyu, K. et al., Omega-3 fatty acids decreased irritability of patients with bipolar disorder in an add-on, open label study. *Nutr. J.* 2005; 4:6

115. Chiu, C.C. et al., Omega-3 fatty acids are more beneficial in the depressive phase than in the manic phase in patients with bipolar I disorder. *J. Clin. Psychiatry* 2005; 66:1613–1614.

116. Marangell, L.B. et al., Omega-3 fatty acids in bipolar disorder: clinical and research considerations. *Prostaglandins Leukotm. Essent. Fatty Acids* 2006; 75:315–321.

117. Rudin, D.O., The major psychoses and neuroses as omega-3 essential fatty acid deficiency syndrome: substrate pellagra. *Biol. Psychiatry* 1981; 16:837–850.

118. Christensen, O., and Christensen, E., Fat consumption and schizophrenia. *Acta Psychiatr. Scand.* 1988; 78:587–591.

119. Assies, J. et al., Significantly reduced docosahexaenoic and docosapentaenoic acid concentrations in erythrocyte membranes from schizophrenic patients compared with a carefully matched control group. *Biol. Psychiatry* 2001; 49:510–522.

120. Khan, M.M. et al., Reduced erythrocyte membrane essential fatty acids and increased lipid peroxides in schizophrenia at the never-medicated first-episode of psychosis and after years of treatment with antipsychotics. *Schizophr. Res.* 2002; 58:1–10.

121. Joy, C.B., Mumby-Croft, R., and Joy, L.A., Polyunsaturated fatty acid supplementation for schizophrenia. *Cochrane Database Syst. Rev.* 2006; 3:CD001257.

122. Markesbery, W.R. et al., Lipid peroxidation is an early event in the brain in amnestic mild cognitive impairment. *Ann. Neurol.* 2005; 58:730–735.
123. Grant, W.B., Dietary links to Alzheimer's Disease. *Alzheimer's Dis. Rev.* 1997; 2:42–55.
124. Morris, M.C. et al., Consumption of fish and n-3 fatty acids and risk of incident Alzheimer Disease. *Arch. Neurol.* 2003; 60:940–946.
125. Schaefer, E.J. et al., Plasma phosphatidylcholine docosahexaenoic acid content and risk of dementia and Alzheimer disease: the Framingham Heart Study. *Arch. Neurol.* 2006; 63:1545–1550.
126. Huang, T.L. et al., Benefits of fatty fish on dementia risk are stronger for those without APOE epsilon4. *Neurology* 2005; 65:1409–1414.
127. Tully, A.M. et al., Low serum cholesteryl ester- docosahexaenoic acids levels in Alzheimer's Disease: a case-control study. *Br. J. Nutr.* 2003; 89:483–490.
128. Soderberg, M., Fatty acid phospholipid composition in ageing and in Alzheimer's Disease. *Lipids* 1991; 26:421–425.
129. Freund-Levi, Y. et al., Omega-3 fatty acid treatment in 174 patients with mild to moderate Alzheimer disease: a randomized double-blind trial. *Arch. Neurol.* 2006; 63:1402–1408.
130. Boston, P.F. et al., Ethyl-EPA in Alzheimer's Disease – a pilot study. *Prostaglandins Leukot. Essent. Fatty Acids* 2004; 71:341–346.
131. Reiffel, J.A., and McDonald, A., Antiarrhythmic effects of omega-3 fatty acids. *Am. J. Cardiol.* 2006; 98:50i–60i.
132. Den Ruijter, H.M. et al., Pro- and antiarrhythmic properties of a diet rich in fish oil. *Cardiovasc. Res.* 2007; 73:316–325.
133. Jabbar, R., and Saldeen, T., A new predictor of risk for sudden cardiac death. *Ups. J. Med. Sci.* 2006; 111:169–177.
134. Mori, T.A., Effect of fish and fish oil-derived omega-3 fatty acids on lipid oxidation. *Redox. Rep.* 2004; 9:193–197.
135. Mozaffarian, D., and Rimm, E.B., Fish intake, contaminants, and human health: evaluating the risks and the benefits. *JAMA* 2006; 296:1885–1899.
136. Foran, S.E., Flood, J.G., and Lewandrowski, K.B., Measurement of mercury levels in concentrated over-the-counter fish oil preparations: is fish oil healthier than fish? *Arch. Pathol. Lab. Med.* 2003; 127:1603–1605.
137. Vuksan, V. et al., Korean red ginseng (Panax ginseng) improves glucose and insulin regulation in well-controlled, type 2 diabetes: results of a randomized, double-blind, placebo-controlled study of efficacy and safety. *Nutr. Metab. Cardiovasc. Dis.* 2006 (Epublished ahead of print).
138. Sotaniemi, E.A., Haapakoski, E., and Rautio, A., Ginseng therapy in non-insulin-dependent diabetic patients. *Diabetes Care* 1995; 18:1373–1375.
139. Tode, T., Kikuchi, Y., Hirata, J., et al., Effect of Korean red ginseng on psychological functions in patients with severe climacteric syndromes. *Int. J. Gynaecol. Obstet.* 1999; 67:169–174.
140. Hasegawa, H. et al., Interactions of ginseng extract, ginseng separated fractions, and some triterpenoid saponins with glucose transporters in sheep erythrocytes. *Planta Med.* 1994; 60:153–157.
141. Waki, I. et al., Effects of a hypoglycemic component of ginseng radix on insulin biosynthesis in normal and diabetic animals. *J. Pharmacobiodyn.* 1982; 5:547–554.
142. Han, K.L. et al., Ginsenoside 20S-protopanaxatriol (PPT) activates peroxisome proliferator-activated receptor gamma (PPARgamma) in 3T3-L1 adipocytes. *Biol. Pharm. Bull.* 2006; 29:110–113.
143. Keum, Y.S. et al., Inhibitory effects of the ginsenoside Rg3 on phorbol ester-induced cyclooxygenase-2 expression, NF-kappaB activation and tumor promotion. *Mutat. Res.* 2003; 523–524:75–85.
144. Cho, J.Y. et al., In vitro inhibitory effect of protopanaxadiol ginsenosides on tumor necrosis factor (TNF)-alpha production and its modulation by known TNF-alpha antagonists. *Planta Med.* 2001; 67:213–218.
145. King, M.L., Adler, S.R., and Murphy, L.L., Extraction-dependent effects of American ginseng (Panax quinquefolium) on human breast cancer cell proliferation and estrogen receptor activation. *Integr. Cancer Ther.* 2006; 5:236–243.

146. Gong, Y.S. and Zhang, J.T., Effect of 17-beta-estradiol and ginsenoside Rg1 on reactive microglia induced by beta-amyloid peptides. *J. Asian Nat. Prod. Res.* 1999; 1:153–161.

147. de Kloet, E.R. et al., Ginsenoside RG1 and corticosteroid receptors in rat brain. *Endocrinol. Jpn.* 1987; 34:213–220.

148. Chen, F., Eckman, E.A., and Eckman, C.B., Reductions in levels of the Alzheimer's amyloid beta peptide after oral administration of ginsenosides. *FASEB J.* 2006; 20:1269–1271.

149. Thommessen, B. and Laake, K., No identifiable effect of ginseng (Gericomplex) as an adjuvant in the treatment of geriatric patients. *Aging (Milano)* 1996; 8:417–420.

150. Caso Marasco, A. et al., Double-blind study of a multivitamin complex supplemented with ginseng extract. *Drugs Exp. Clin. Res.* 1996; 22:323–329.

151. Cota. D. et al., Hypothalamic mTOR signaling regulates food intake. *Science* 2006; 312:927–930.

152. Roh. C. et al., Nutrient-sensing mTOR-mediated pathway regulates leptin production in isolated rat adipocytes. *Am. J. Physiol. Endocrinol. Metab.* 2003; 284:E322–330.

153. Lynch, C.J. et al., Leucine in food mediates some of the postprandial rise in plasma leptin concentrations. *Am. J. Physiol. Endocrinol. Metab.* 2006; 291:621–630.

154. Layman, D.K. and Baum, J.I., Dietary protein impact on glycemic control during weight loss. *J. Nutr.* 2004; 134:968–973.

155. Petroulakis, E. et al., mTOR signaling: implications for cancer and anticancer therapy. *Br. J. Cancer* 2006; 94:195–199.

156. Lang, C.H., Deshpande, N., and Frost, R.A., Leucine acutely reverses burn-induced alterations in translation initiation in heart. *Shock* 2004; 22:326–332.

157. van der Mast, R.C. and Fekkes, D., Serotonin and amino acids: partners in delirium pathophysiology? *Semin. Clin. Neuropsychiatry* 2000; 5:125–131.

158. Delgado, P.L. et al., Serotonin and the neurobiology of depression. Effects of tryptophan depletion in drug-free depressed patients. *Arch. Gen. Psychiatry* 1994; 51:865–874.

159. Hutson, S.M., Sweatt, A.J., and Lanoue, K.F., Branched-chain [corrected] amino acid metabolism: implications for establishing safe intakes. *J. Nutr.* 2005; 135(6 Suppl.): 1557S–1564S.

160. Zemel, M.B., The role of dairy foods in weight management. *J. Am. Coll. Nutr.* 2005; 24(6 Suppl.):537S–546S.

161. Park, Y.K. et al., Effectiveness of food fortification in the United States: the case of pellagra. *Am. J. Public Health.* 2000; 90:727–738.

162. Altschul, R., Hoffer, A., and Stephen, J.D., Influence of nicotinic acid on serum cholesterol in man. *Arch. Biochem. Biophys.* 1955; 54:558–559.

163. Knopp, R.H. et al., Equivalent efficacy of a time-release form of niacin (Niaspan) given once-a-night versus plain niacin in the management of hyperlipidemia. *Metabolism* 1998; 47:1097–1104.

164. Shepherd, J. et al., Effects of nicotinic acid therapy on plasma high density lipoprotein subfraction distribution and composition and on apolipoprotein A metabolism. *J. Clin. Invest.* 1979; 63:858–867.

165. Grundy, S.M. et al., Influence of nicotinic acid on metabolism of cholesterol and triglycerides in man. *J. Lipid. Res.* 1981; 22:24–36.

166. Cheraskin, E., Ringsdorf, W.M., and Medford, F.H., The "ideal" daily niacin intake. *Int. J. Vitamin Nutr. Res.* 1976; 46:58–60.

167. Wise, A. et al., Molecular identification of high and low affinity receptors for nicotinic acid. *J. Biol. Chem.* 2003; 278:9869–9874.

168. Knowles, H.J. et al., Niacin induces PPARgamma expression and transcriptional activation in macrophages via HM74 and HM74a-mediated induction of prostaglandin synthesis pathways. *Biochem. Pharmacol.* 2006; 71:646–656.

169. Parodi, A., Guarrera, M., and Rebora, A., Nicotinic acid flush. *Lancet* 1981; 2:477.

170. Clementz, G.L., and Holmes, A.W., Nicotinic acid-induced fulminant hepatic failure. *J. Clin. Gastroenterol.* 1987; 9:582–584.

171. Farmer, J.A., and Gotto, A.M. Jr. Antihyperlipidaemic agents. Drug interactions of clinical significance. *Drug Saf.* 1994; 11:301–309.

172. Kelly, J.J. et al., Effects of nicotinic acid on insulin sensitivity and blood pressure in healthy subjects. *J. Hum. Hypertens.* 2000; 14:567–572.

173. Canner, P.L., Furberg, C.D., and McGovern, M.E., Benefits of niacin in patients with versus without the metabolic syndrome and healed myocardial infarction (from the Coronary Drug Project). *Am. J. Cardiol.* 2006; 97:477–479.

174. Shepherd, J. et al., Nicotinic acid in the management of dyslipidaemia associated with diabetes and metabolic syndrome: a position paper developed by a European Consensus Panel. *Curr. Med. Res. Opin.* 2005; 21:665–682.

175. Ban, T.A., Nicotinic acid in the treatment of schizophrenias. Practical and theoretical considerations. *Neuropsychobiology* 1975; 1:133–145.

176. Smythies, J., The adrenochrome hypothesis of schizophrenia revisited. *Neurotox. Res.* 2002; 4:147–150.

177. Horrobin, D.F., Schizophrenia: a biochemical disorder? *Biomedicine* 1980; 32:54–55.

178. Lin, S.H. et al., Familial aggregation in skin flush response to niacin patch among schizophrenic patients and their nonpsychotic relatives. *Schizophr. Bull.* 2007; 33:174–182.

179. Berger, G.E., Smesny, S., and Amminger, G.P., Bioactive lipids in schizophrenia. *Int. Rev. Psychiatry* 2006; 18:85–98.

180. Wang, Z. et al., Dealcoholized red wine containing known amounts of resveratrol suppresses atherosclerosis in hypercholesterolemic rabbits without affecting plasma lipid levels. *Int. J. Mol. Med.* 2005; 16:533–540.

181. Miatello, R. et al., Chronic administration of resveratrol prevents biochemical cardiovascular changes in fructose-fed rats. *Am. J. Hypertens.* 2005; 18:864–870.

182. Su, H.C. et al., Resveratrol, a red wine antioxidant, possesses an insulin-like effect in streptozotocin-induced diabetic rats. *Am. J. Physiol. Endocrinol. Metab.* 2006; 290:339–346.

183. Bastianetto, S. et al., Neuroprotective effects of resveratrol against beta-amyloid-induced neurotoxicity in rat hippocampal neurons: involvement of protein kinase C. *Br. J. Pharmacol.* 2004; 141:997–1005.

184. Bertelli, A.A. et al., Resveratrol inhibits TNF alpha-induced endothelial cell activation. *Therapie* 2001; 56:613–616.

185. Frankel, E.N. et al., Inhibition of oxidation of human low-density lipoprotein by phenolic substances in red wine. *Lancet* 1993; 341:454–457.

186. Howitz, K.T. et al., Small molecule activators of sirtuins extend Saccharomyces cerevisiae lifespan. *Nature* 2003; 425:191–196.

187. Valenzano, D.R. et al., Resveratrol prolongs life span and retards the onset of age-related markers in a short-lived vertebrate. *Curr. Biol.* 2006; 16:296–300.

188. Walle, T. et al., High absorption but very low bioavailability of oral resveratrol in humans. *Drug Metab. Dispos.* 2004; 32:1377–1382.

189. Zern, T.L. et al., Grape polyphenols exert a cardioprotective effect in pre- and postmenopausal women by lowering plasma lipids and reducing oxidative stress. *J. Nutr.* 2005; 135:1911–1917.

190. Gehm, H. et al., Resveratrol, a polyphenolic compound found in grapes and wine, is an agonist for the estrogen receptor. *Proc. Nat. Acad. Sci.* 1997; 94:557–562.

191. Cantoni, G.L., Activation of methionine for transmethylation. *J. Biol. Chem.* 1951; 189:745–754.

192. Alvarez, L. et al., Analysis of the 5′ non-coding region of rat liver S-adenosylmethionine synthetase mRNA and comparison of the Mr deduced from the cDNA sequence and the purified enzyme. *FEBS Lett.* 1991; 290; 142–146.

193. Cankurtaran, M. et al., Prevalence and correlates of metabolic syndrome (MS) in older adults. *Arch. Gerontol. Geriatr.* 2006; 42:35–45.

194. Neeman, G. et al., Relation of plasma glycine, serine, and homocysteine levels to schizophrenia symptoms and medication type. *Am. J. Psychiatry* 2005; 162:1738–1740.

195. Mantero, M. et al., Controlled double-blind study (SAMe-imipramine) in depressive syndrome. *Minerva Med.* 1975; 66:4098–4101.

196. Williams, A.L. et al., S-adenosylmethionine (SAMe) as treatment for depression: a systematic review. *Clin. Invest. Med.* 2005; 28:132–139.

197. Bottiglieri, T. and Hyland, K., S-adenosylmethionine levels in psychiatric and neurological disorders: a review. *Acta Neurol. Scand.* 1994; 154(Suppl.):19–26.

198. Bell, K.M., S-adenosylmethionine blood levels in major depression: changes with drug treatment. *Acta Neurol. Scand.* 1994; 154(Suppl.):15–18.

199. Hickie, I. et al., Late-onset depression: genetic, vascular and clinical contributions. *Psychol. Med.* 2001; 31:1403–1412.

200. Charlton, C.G. and Crowell, B. Jr., Striatal dopamine depletion, tremors, and hypokinesia following the intracranial injection of S-adenosylmethionine: a possible role of hypermethylation in parkinsonism. *Mol. Chem. Neuropathol.* 1995; 26:269–284.

201. Otero-Losada, M.E. and Rubio, M.C., Acute changes in 5-HT metabolism after S-adenosyl-L-methionine administration. *Gen. Pharmacol.* 1989; 20:403–406.

202. Mann, S.P. and Hill, M.W., Activation and inactivation of striatal tyrosine hydroxylase: the effects of pH, ATP and cyclic AMP, S-adenosylmethionine and S-adenosylhomocysteine. *Biochem. Pharmacol.* 1983; 32:3369–3374.

203. Foxton, R.H., Land, J.M., and Heales, S.J. Tetrahydrobiopterin availability in Parkinson's and Alzheimer's Disease; potential pathogenic mechanisms. *Neurochem. Res.* 2007; 32:751–756.

204. Morrison, L.D., Becker, L., and Kish, S.J., S-adenosylmethionine decarboxylase in human brain. Regional distribution and influence of aging. *Brain Res. Dev. Brain Res.* 1993; 73:237–241.

205. Genedani, S. et al., Influence of SAMe on the modifications of brain polyamine levels in an animal model of depression. *Neuroreport.* 2001; 12:3939–3942.

206. Zomkowski, A.D., Santos, A.R., and Rodrigues, A.L., Putrescine produces antidepressant-like effects in the forced swimming test and in the tail suspension test in mice. *Prog. Neuropsychopharmacol. Biol. Psychiatry* 2006; 30:1419–1425.

207. Morrison, L.D., Smith, D.D., and Kish, S.J., Brain s-adenosylmethionine levels are severely decreased in Alzheimer's Disease. *J. Neurochem.* 1996; 67:1328–1331.

208. Morrison, L.D., Bergeron, C., and Kish, S.J., Brain S-adenosylmethionine decarboxylase activity is increased in Alzheimer's Disease. *Neurosci. Lett.* 1993; 154:141–144.

209. Blusztajn, J.K., Zeisel, S.H., and Wurtman, R.J., Synthesis of lecithin (phosphatidylcholine) from phosphatidylethanolamine in bovine brain. *Brain Res.* 1979; 179:319–327.

210. Scarpa, S. et al., Gene silencing through methylation: an epigenetic intervention on Alzheimer' disease. *J. Alzheimer's Dis.* 2006; 9:407–414.

211. Sontag, E. et al., Protein phosphatase 2A methyltransferase links homocysteine metabolism with tau and amyloid precursor protein regulation. *J. Neurosci.* 2007; 27:2751–2759.

212. Scarpa, S. et al., Presenilin 1 gene silencing by S-adenosylmethionine: a treatment for Alzheimer disease? *FEBS Lett.* 2003; 541:145–148.

213. Cohen, B.M., Satlin, A., and Zubenko, G.S., S-adenosyl-L-methionine in the treatment of Alzheimer's Disease. *J. Clin. Psychopharmacol.* 1988; 8:43–47.

214. Mulder, C. et al., The transmethylation cycle in the brain of Alzheimer patients. *Neurosci. Lett.* 2005; 386:69–71.

215. Setola, E. et al., Insulin resistance and endothelial function are improved after folate and vitamin B12 therapy in patients with metabolic syndrome: relationship between homocysteine levels and hyperinsulinemia. *Eur. J. Endocrinol.* 2004; 151:483–489.

216. Jin. C.J. et al., S-adenosyl-L-methionine increases skeletal muscle mitochondrial DNA density and whole body insulin sensitivity in OLETF rats. *J. Nutr.* 2007; 137:339–344.

217. Boger, R.H. et al., LDL cholesterol upregulates synthesis of asymmetrical dimethylarginine in human endothelial cells: involvement of S-adenosylmethionine-dependent methyltransferases. *Circ. Res.* 2000; 87:99–105.

218. Tessari, P. et al., Effects of insulin on methionine and homocysteine kinetics in Type 2 diabetics with neuropathy. *Diabetes* 2005; 54:2968–2976.

219. Cave, M. et al., Nonalcoholic fatty liver disease: predisposing factors and the role of nutrition. *J. Nutr. Biochem.* 2007; 18:184–195.

220. Mato, J.M. et al., S-adenosylmethionine: a control switch that regulates liver function *FASEB J.* 2002; 16:15–26

221. Lu, S.C. et al., Methionine adenosyltransferase 1A knockout mice are predisposed to liver injury and exhibit increased expression of genes involved in proliferation. *Proc. Natl. Acad. Sci. U.S.A.* 2001; 98:5560–5565.

222. Yu, J., Sauter, S., and Parlesak, A., Suppression of TNF-alpha production by S-adenosylmethionine in human mononuclear leukocytes is not mediated by polyamines. *Biol. Chem.* 2006; 387:1619–1627.

223. Veal, N. et al., Inhibition of lipopolysaccharide-stimulated TNF-alpha promoter activity by S-adenosylmethionine and 5'-methylthioadenosine. *Am. J. Physiol. Gastrointest. Liver Physiol.* 2004; 287:G352–362.

224. Chang, C.Y. et al., Therapy of NAFLD: antioxidants and cytoprotective agents. *J. Clin. Gastroenterol.* 2006; 40(Suppl. 1):S51–60.

225. Goren, J.L. et al., Bioavailability and lack of toxicity of S-adenosyl-L-methionine (SAMe) in humans. *Pharmacotherapy* 2004; 24:1501–1507.

226. Angulo, P., Current best treatment for non-alcoholic fatty liver disease. *Expert Opin. Pharmacother.* 2003; 4:611–623.

227. Kren, V. and Walterová, D., Silybin and silymarin – new effects and applications. *Biomed. Pap. Med. Fac. Univ. Palacky Olomouc Czech Repub.* 2005; 149:29–41.

228. Rainone, F., Milk thistle. *Am. Fam. Physician.* 2005; 72:1285–1288.

229. Manna, S.K. et al., Silymarin suppresses TNF-induced activation of NF-kappa B, c-Jun N-terminal kinase, and apoptosis. *J. Immunol.* 1999; 163:6800–6809.

230. Permana, P.A., Menge, C., and Reaven, P.D., Macrophage-secreted factors induce adipocyte inflammation and insulin resistance. *Biochem. Biophys. Res. Commun.* 2006; 341:507–514.

231. Velussi, M. et al., Long-term (12 months) treatment with an anti-oxidant drug (silymarin) is effective on hyperinsulinemia, exogenous insulin need and malondialdehyde levels in cirrhotic diabetic patients. *J. Hepatol.* 1997; 26:871–879.

232. Soto, C. et al., Silymarin induces recovery of pancreatic function after alloxan damage in rats. *Life Sci.* 2004; 75:2167–2180.

233. Somogyi, A. et al., Short term treatment of type II hyperlipoproteinaemia with silymarin. *Acta Med. Hung.* 1989; 46:289–295.

234. Skottová, N. et al., Effects of polyphenolic fraction of silymarin on lipoprotein profile in rats fed cholesterol-rich diets. *Pharmacol. Res.* 2003; 47:17–26.

235. Skottová, N. et al., Phenolics-rich extracts from Silybum marianum and Prunella vulgaris reduce a high-sucrose diet induced oxidative stress in hereditary hypertriglyceridemic rats. *Pharmacol. Res.* 2004; 50:123–130.

236. Locher, R. et al., Inhibitory action of silibinin on low density lipoprotein oxidation. *Arzneimittelforschung* 1998; 48:236–239.

237. Rui, Y.C. et al., Effects of silybin on production of oxygen free radical, lipoperoxide and leukotrienes in brain following ischemia and reperfusion. *Zhongguo Yao Li Xue Bao* 1990; 11:418–421.

238. Nielsen, F.H., Nutritional requirements for boron, silicon, vanadium, nickel, and arsenic: current knowledge and speculation. *FASEB J.* 1991; 5:2661–2667.

239. Marzban, L. and McNeill, J.H., Insulin-like actions of vanadium: potential as a therapeutic agent. *J. Trace Elem. Exp. Med.* 2003; 16:253–267.

240. Jentjens, R.L. and Jeukendrup, A.E., Effect of acute and short-term administration of vanadyl sulphate on insulin sensitivity in healthy active humans. *Int. J. Sport Nutr. Exerc. Metab.* 2002; 12:470–479.

241. Li, J. et al., Antilipolytic actions of vanadate and insulin in rat adipocytes mediated by distinctly different mechanisms. *Endocrinology* 1997; 138:2274–2279.

242. Wang, J. et al., Effect of vanadium on insulin sensitivity and appetite. *Metabolism* 2001; 50:667–673.

243. Nagai, M. et al., Pentavalent vanadium at concentration of the underground water level enhances the sweet taste sense to glucose in college students. *Biometals* 2006; 19:7–12.

244. Poucheret, P. et al., Vanadium and diabetes. *Mol. Cell. Biochem.* 1998; 188:73–80.

245. Cohen, N. et al., Oral vanadyl sulphate improves hepatic and peripheral insulin sensitivity in patients with non-insulin dependent Diabetes Mellitus. *J. Clin. Invest.* 1995; 95:2501–2509.

246. Goldfine, A.B. et al., Metabolic effects of vanadyl sulfate in humans with non-insulin dependent diabetes mellitus: in vivo and in vitro studies. *Metabolism* 2000; 49:400–410.

247. Curran, G.L. et al., Effect of cholesterol synthesis inhibition in normocholesteremic young men. *J. Clin. Invest.* 1959; 38:1251–1261.

248. Naylor, G.J., Vanadium and manic depressive psychosis. *Nutr. Health* 1984; 3:79–85.

249. Norman, A.W. et al., Vitamin D deficiency inhibits pancreatic secretion of insulin. *Science* 1980; 209:823–825.

250. Gedik, O. and Akalin, S., Effects of vitamin D deficiency and repletion on insulin and glucagon secretion in man. *Diabetologia* 1986; 29:142–145.

251. Kumar, S. et al., Improvement in glucose tolerance and beta-cell function in a patient with vitamin D deficiency during treatment with vitamin D. *Postgrad. Med. J.* 1994; 70:440–443.

252. Isaia, G., Giorgino, R., and Adami, S., High prevalence of hypovitaminosis D in female type 2 diabetic population. *Diabetes Care* 2001; 24:1496.

253. Mathieu, C. et al., Vitamin D and diabetes. *Diabetologia* 2005; 48:1247–1257.

254. Chiu, K.C. et al., Hypovitaminosis D is associated with insulin resistance and beta cell dysfunction. *Am. J. Clin. Nutr.* 2004; 79:820–825.

255. Ford, E.S. et al., Concentrations of serum vitamin D and metabolic syndrome among U.S. adults. *Diabetes Care* 2005; 28:1228–1230.

256. Sowers, M.R., Wallace, R.B., and Lemke, J.H., The association of intakes of vitamin D and calcium with blood pressure among women. *Am. J. Clin. Nutr.* 185; 42:135–142.

257. Forman, J.P. et al., Vitamin D intake and risk of incident hypertension: results from three large prospective cohort studies. *Hypertension* 2005; 46:676–682.

258. Walbert, T., Jirikowski, G.F., and Prufer, K. Distribution of 1,25-dihydroxyvitamin D3 receptor immunoreactivity in the limbic system of the rat. *Horm. Metab. Res.* 2001; 33:525–531.

259. Gloth, F.M. 3rd, Alam, W., and Hollis, B., Vitamin D vs broad spectrum phototherapy in the treatment of seasonal affective disorder. *J. Nutr. Health Aging* 1999; 3:5–7.

260. Lansdowne, A.T., and Provost, S.C., Vitamin D3 enhances mood in healthy subjects during winter. *Psychopharmacology (Berl.)* 1998; 135:319–323.

261. Schneider, B. et al., Vitamin D in schizophrenia, major depression and alcoholism. *J. Neural. Transm.* 2000; 107:839–842.

262. Sato, Y., Asoh, T., and Oizumi, K., High prevalence of vitamin D deficiency and reduced bone mass in elderly women with Alzheimer's Disease. *Bone* 1998; 23:555–557.

263. Sutherland, M.K. et al., Reduction of vitamin D hormone receptor mRNA levels in Alzheimer as compared to Huntington hippocampus: correlation with calbindin-28k mRNA levels. *Brain Res. Mol. Brain Res.* 1992; 13:239–250.

264. Obradovic, D. et al., Cross-talk of vitamin D and glucocorticoids in hippocampal cells. *J. Neurochem.* 2006; 96:500–509.

265. Sonnenberg, J. et al., 1,25-Dihydroxyvitamin D3 treatment results in increased choline acetyltransferase activity in specific brain nuclei. *Endocrinology* 1986; 118:1433–1439.

266. Holick, M.F., Vitamin D: The underappreciated D-lightful hormone that is important for skeletal and cellular health. *Curr. Opin. Endocrinol. Diabetes* 2002; 9:87–98.

267. Matsuoka, L.Y., Wortsman, J., and Hollis, B.W., Use of topical sunscreen for the evaluation of regional synthesis of vitamin D3. *J. Am. Acad. Dermatol.* 1990; 22:772–775.

268. Tangpricha, V. et al., Vitamin D insufficiency among free-living healthy young adults. *Am. J. Med.* 2002; 112:659–662.

269. Holick, M.F., High prevalence of vitamin D inadequacy and implications for health. *Mayo. Clin. Proc.* 2006; 81:353–373.

⧠⧠ CONCLUSION: METABOLIC SYNDROME AND WHAT TO DO ABOUT IT

The Metabolic Syndrome is a condition to which many individuals are predisposed through inheritence of low insulin sensitivity. However, the syndrome is exacerbated by comorbidities and driven by unwise choices in diet and lifestyle. The incidence of Metabolic Syndrome has exploded to epidemic proportions in the United States. Initially, it was seen as a constellation of metabolic changes that increased the risk for diabetes and cardiovascular disease. However, in the years since Dr. Gerald Reaven first described the condition, under the name of Syndrome X, our understanding of Metabolic Syndrome has grown enormously. It is now known that Metabolic Syndrome predisposes its sufferers not only to diabetes and cardiovascular disease, but to a panoply of other ailments that includes Alzheimer's Disease, sleep disorders, sexual disturbances, Major Depression, Schizophrenia and other forms of psychiatric illness.

Although Metabolic Syndrome involves enormously complicated physiology, there are simple and straightforward ways to deal with it. Effectively addressing Metabolic Syndrome requires long-term changes in diet, exercise, sleep hygiene, elimination of bad habits such as smoking, and stress reduction. Of particular importance is the diagnosis and treatment of psychiatric illnesses that can precipitate and worsen Metabolic Syndrome. In some cases, medications may be necessary to treat Metabolic Syndrome and comorbid conditions. In other cases, changes in medications may be the most prudent step.

THE FIRST STEP

The first step in diagnosing, treating, or simply avoiding Metabolic Syndrome is a thorough medical evaluation. A good physical examination and comprehensive set of laboratory studies are required to determine the degree to which a patient meets the criteria for Metabolic Syndrome. The criteria for Metabolic Syndrome proposed by the Adult Treatment Panel of the National Cholesterol Education Program (ATP III) have probably been the most widely accepted ones. A diagnosis of Metabolic Syndrome can be made when a patient exhibits three or more of the following criteria: Abdominal obesity, as measured by waist circumference over 40 inches for men or 35 inches for women; a fasting blood glucose level over 110 mg/dl; blood pressures greater than 130/80; serum triglyceride levels over 150 mg/dl; or serum HDL levels under 40 mg/dl for men or 50 mg/dl for women.

It is worth noting that the criteria of the ATP III only suggest the likelihood of insulin resistance. Meeting these criteria does not entail the existence of insulin resistance, nor does failure to meet the criteria rule out insulin resistance. Dr. Reaven has emphasized the fact that insulin resistance can be pushing a patient into pathology without necessarily meeting the criteria for the Metabolic Syndrome[1]. The most important component of Metabolic Syndrome has always been insulin resistance, which is not one of the ATP III criteria for diagnosis of the syndrome. This is not out of negligence on the part of the medical community. Rather it is due to the fact that direct measurements of insulin resistance are difficult. Fortunately, Reaven and his coworkers have found that some relatively simple laboratory tests may be just as sensitive as the ATP III criteria in screening for Metabolic Syndrome, and possibly more accurate in indicating insulin resistance. A ratio of triglyceride to HDL over 3, particularly in the presence of fasting insulin over 109 pmol/l is strongly suggestive of insulin resistance[2]. In lieu of the elaborate research technique of precisely measuring insulin resistance, obtaining these values, along with assessing measures as per ATP III criteria, is probably the best clinical approach to recognize this component of Metabolic Syndrome.

Assessing the degree of Metabolic Syndrome and insulin resistance in a patient will help establish that patient's need to lose weight and reduce risk factors for cardiovascular disease, diabetes, Alzheimer's Disease and psychiatric illness. Just as importantly, it will help determine the best method for doing so.

DIET

In individuals who are found to have Metabolic Syndrome and obesity, it will be prudent to start them on a low-carbohydrate diet for both weight loss and reduction in demand for insulin. There may be preexisting conditions, such as gout or a history of renal stones, that are contraindications for high protein intake. Otherwise, if weight loss is a major consideration, a diet high in protein and fat is reasonable. I cannot overemphasize the prudence of modifying the otherwise troubling Atkin's type of diet. This modification is the minimizing of saturated fat, while optimizing EPA, DHA and monounsaturated fat. This can be as simple as replacing beef, pork and other meats high in saturated fat with fish and game meats. Egg substitute or *omega-3* fatty acid-enriched eggs are excellent additions to the diet. Olive oil should be used when possible and replace butter in food preparation. Margarines without *trans* fatty acids but rich in polyunsaturated fats and *omega-3* fatty acids are also worthwhile substitutes for butter.

While on a high-fat and protein, low-carbohydrate diet, it is imperative that the intake of certain nutrients lessened by the reduction of carbohydrate intake be restored to beneficial levels. There is no compelling reason that supplements cannot provide perfectly adequate sources of these nutrients. A good, potent multivitamin should be taken. Because of the extra oxidation of fat, it is prudent to add vitamins E and C. Chromium is likely to be lost if plant sources of food are reduced. Because of chromium's ability to enhance insulin's effects, this essential mineral is particularly useful in treating Metabolic Syndrome. Adequate intake of calcium, magnesium and zinc are necessary to optimize insulin sensitivity. Fiber is an important addition, as well as lots of pure water.

The benefits of the high-protein and fat diet may last only a few months. At some point, switching to a more conservative diet is advisable. The Mediterranean style diets are excellent. They have been shown to improve symptoms of Metabolic Syndrome, reduce insulin resistance, and reverse some of the sequelae of Metabolic Syndrome. Aside from the hype, the South Beach, Perricone and Rosedale Diets have many benefits to offer. The less adventurous will do well on Reaven's Syndrome X Diet.

Some individuals are blessed with high sensitivity to insulin. They are unlikely to develop Metabolic Syndrome, and they will obviously be at far less risk for the maladies that evolve out of insulin resistance. However, these people are not immune from obesity, hyperlipidemia, hypercholesterolemia and other risk factors for serious illness. If they are obese, they can lose weight with reduced calorie low-fat, high-carbohydrate diets. They will not suffer the increases in triglycerides and decreases in HDL that people with insulin resistance are prone to developing on such diets.

There are some foods and ingredients that everyone should stay away from, regardless of whether or not they have insulin resistance and Metabolic Syndrome. Carbohydrates with high glycemic indexes should avoided. Although people with good insulin sensitivity may tolerate high loads of carbohydrate, they are not immune from the postprandial hypoglycemia that may result from the consumption of sugary treats. High concentrations of fructose, such as is found in high-fructose corn syrup, should particularly be avoided. In fact, high-fructose diets are often used in studies to induce Metabolic Syndrome in experimental animals. If an individual was born with little likelihood of developing Metabolic Syndrome, dietary abuses of fructose and high glycemic index carbohydrates will improve their chances. *Trans* fatty acids are unnatural, toxic, and without redeeming features. Eliminate them. This means watching out for hydrogenated oils often found in processed foods, as well as curtailing visits to many fast-food restaurants.

EXERCISE

Obesity exacerbates Metabolic Syndrome. It also contributes independently to cardiovascular disease and other health problems. It can diminish quality of life and complicate psychiatric illness and its treatment. Exercise burns calories, which, in turn, contributes to weight loss. For this reason alone, increasing physical activity is helpful in the treatment of Metabolic Syndrome. However, the effects of exercise go well beyond simply helping lose weight. Low cardiorespiratory fitness is a strong and independent predictor of Metabolic Syndrome. Cardiovascular fitness is inversely related to the presence of Metabolic Syndrome, and this is true regardless of an individual's weight[3]. Simple physical inactivity contributes to the development of Metabolic Syndrome[4]. An extra 3 hours a day of sitting in front of the television has been found, over the long run, to independently erase the benefits that could be gained by participating for 2 hours a week in vigorous physical activity or adding 12 g of fiber per day to the diet[5]. On the other hand, initiating a program of exercise can improve the physiological measures of Metabolic Syndrome[6]. Tell your patients to "get up, get out, and get going". The simple types of aerobic exercises that can improve Metabolic Syndrome include walking briskly for 30 minutes, stairwalking for 15 minutes, riding a bicycle 5 miles in 30 minutes, or dancing for 30 minutes[7]. It is not even necessary that such exercise be done all in one session. The activity can be split up into several shorter sessions during the day without significant loss of benefits. While the cardio-respiratory improvements of aerobic training are helpful, adding weight training makes it even better[8]. Along with helping to reverse Metabolic Syndrome, exercise is also an excellent augmentation strategy in the treatment of Major Depression[9] and schizophrenia[10]. Exercise can also be a helpful addition to the treatment of fibromyalgia[11]. The addition of exercise to any therapeutic program will provide benefits.

SLEEP

Lack of sleep, broken sleep, erratic sleep schedules, and sleep complicated by apnea all contribute to Metabolic Syndrome. These sleep problems also aggravate psychiatric illnesses. I have often found that the single most important step I can take in treating an affective or psychotic disorder is to give my patient consistently good, full nights of sleep. If a point of crisis in sleep loss has been reached and medication for sleep is needed, than by all means use it. There are few sleep medications more dangerous than not sleeping.

Many patients begin to sleep better when their primary psychiatric illness is brought under control. Still, it is important to assess the degree to which shift work, late work nights, and burning the candle at both ends can be addressed to allow a more healthy sleep pattern. Many patients, particularly those with BPAD, should not work at night. They should have 9 a.m. to 5 p.m. jobs.

Patients should also be educated in the basic principles of sleep hygiene. Sleep schedules should remain consistent, even on weekends. Have a cool, dark, comfortable and quiet sleeping environment. Use the bed for sleep and sex only. Do not eat or exercise within 2 or 3 hours before bedtime. These steps, along with the avoidance of stimulants such as coffee and cigarettes, should help many patients acquire a healthy sleep schedule.

Sleep apnea is an increasingly common problem that contributes to Metabolic Syndrome, cardiovascular disease and psychiatric illness. Unfortunately, it is far too often overlooked. Many doctors see it as a problem only when it begins to compromise cardiac and pulmonary function. I have seen many cases of individuals diagnosed with Major Depression, Bipolar Disorder, psychosis and dementia who are simply suffering from the loss of sleep, hypoxia and stress of sleep apnea. It is often necessary to treat the sleep apnea with CPAP or other pressured breathing devices. In some individuals, devices placed in the mouth to push the lower jaw forward can be extremely effective. In many cases, the weight loss required to deal with a patient's Metabolic Syndrome can also help them resolve their sleep apnea. However, not everyone with sleep apnea is obese, nor does weight loss always resolve the problem.

Because sleep apnea tends to aggravate Metabolic Syndrome and psychiatric conditions, being active in diagnosing and treating the sleep apnea will further the treatment of those other conditions. I suspect sleep apnea in any patient who reports daytime mental fatigue and a lack of restorative sleep at night. The likelihood of sleep apnea increases if the patient snores and suffers obesity. Although the most definitive diagnosis of sleep apnea is made in the sleep laboratory, I have found that simple, sleep oximetry studies performed at home in the patient's own bed can be extremely useful in screening for sleep apnea. In fact, recent studies show that the expensive and inconvenient polysomnography evaluations in sleep laboratories have no particular advantage over in-home sleep oximetry studies in diagnosisng sleep apnea[12].

STRESS REDUCTION

A high level of serum cortisol is often a shared feature of Metabolic Syndrome and Major Depression. Cortisol antagonizes the effects of insulin and accelerates deposition of visceral fat. These and other physiological effects of cortisol feed the Metabolic Syndrome. The Metabolic Syndrome in turn only worsens psychiatric illness. Cortisol itself exacerbates Major Depression and plays a role in the progression of Alzheimer's Disease and likely other forms of dementia. Stress and the concomitant release of the stress hormone cortisol must be reduced.

There are many techniques to relieve stress and reduce levels of cortisol. Reductions in serum or salivary cortisol have been reported during listening to music[13], meditation[14], yoga[15], Tai Chi[16], or progressive relaxation[17]. People with spiritual convictions, derived from organized religions or not, have lower levels of cortisol than those who do not. Laughter reduces cortisol[18]. People should enjoy themselves, tell jokes, and watch funny movies. Live a little! However, it is a simple fact that in some cases, psychiatric problems will need to be resolved before stress can be meaningfully reduced. This may involve counseling, medication, or both. It is my experience that a significant number of people will not do well without such intervention.

BAD HABITS

Bad habits of many kinds contribute to Metabolic Syndrome. Smoking increases insulin resistance, LDL, triglycerides, fibrinogen, CRP, and homocysteine. It decreases HDL. It exacerbates Metabolic Syndrome. Although people often gain weight when they stop smoking, their lipid profiles are still improved by quitting. I always ask my patients if they would like to quit smoking, and if they want help to do it. A little alcohol may be good, but too much is not. A glass of wine or a beer every other day is probably the reasonable limit. A patient that haggles over such limitations tends to raise my suspicions of how much alcohol they actually consume. There are some patients whose history of abuse of alcohol dictates that they leave it alone. Many patients with psychiatric illness, particularly those with BPAD, like to smoke marijuana. I do not doubt that it can have therapeutic effects in many patients. However, the infamous "marijuana munchies" contribute to obesity and Metabolic Syndrome. Do not be afraid to ask about marijuana use, and be willing to replace the bedtime "joint" with something else to, at least temporarily, help lull patients into the sleep that marijuana helps them achieve. Guzzling cans of high-fructose soda all day, and sitting for hours in front of the TV nibbling chips or popcorn may also need to be addressed.

TREAT CO-MORBID ILLNESS

Four of the major psychiatric illnesses, Major Depression, BPAD, Anxiety Disorders, and Schizophrenia, are associated with a high incidence of Metabolic Syndrome. In each of these illnesses, there may be underlying pathophysiology that predisposes patients to both their psychiatric illness and Metabolic Syndrome. However, as has been shown in path analyses of patients with Major Depression, it is often the case that psychiatric illness progresses into Metabolic Syndrome. This may be due to poor health maintenance, disrupted sleep, carbohydrate craving, bad dietary habits, smoking, and lack of exercise. Effective treatment of psychiatric illness will lessen the likelihood of progression to Metabolic Syndrome. Keep in mind that the appearance of Metabolic Syndrome in the course of psychiatric illness may well herald a point at which both the psychiatric illness and the Metabolic Syndrome will be more difficult to control.

REMEMBER HIPPOCRATES

In some cases, treating Metabolic Syndrome amounts to undoing what we doctors have done to our patients. Well-meaning attempts to treat psychiatric illness may worsen Metabolic Syndrome. Certainly, psychiatric medications are not the only ones that cause Metabolic Syndrome. For example, the acne medication Accutane has been reported to cause hypertriglyceridemia, hypercholesterolemia, obesity and insulin resistance[19] in some patients. Some protease inhibitors used for the treatment of HIV may also cause insulin resistance and other components of Metabolic Syndrome[20]. Nonetheless, psychiatric medications, as well as some antiepileptic medications borrowed by psychiatry for use as mood stabilizers, seem to be the major culprits. We must not overlook the opportunity to replace medications that adversely affect Metabolic Syndrome. Unfortunately, there are cases in which the only effective medication will exacerbate Metabolic Syndrome.

Changes in diet and lifestyle, and addition of certain supplements, as I have discussed in the section on such supplements, can be very helpful in controlling Metabolic Syndrome in patients receiving atypical antipsychotics and other psychiatric medications. However, such measures are not always effective. The consensus opinion is to closely monitor these patients. If changes in diet and lifestyle do not bring Metabolic Syndrome under control after 3 months, then it is time to either try a different medication or to initiate pharmacological treatment to counter those adverse effects. It may be necessary to treat patients with insulin sensitizing medications, antihypertensives, statins or other medications. By and large, there are no contraindications for using standard medications to treat hypertension, hyperlipidemia and hyperglycemia when psychiatric medications cause Metabolic Syndrome. There is still uncertainty about which medications if any should be added to reverse weight gain caused by psychiatric medicines such as atypical antipsychotics. A great deal of attention is being focused upon this problem and, hopefully, satisfactory solutions will be forthcoming. The same approach holds true for patients that develop Metabolic Syndrome due to genetic predisposition or imprudent diet. The patient should be advised to follow a healthier diet and lifestyle, and a reasonable amount of time, perhaps 3 months, should be given to see if these measures are effective. It may also be useful to add one or more dietary supplements to augment those benefits. However, the effects of Metabolic Syndrome are cumulative and dangerous. It is imprudent to delay treatment with medications for hypertension, hyperlipidemia and hyperglycemia if diet and lifestyle measures are ineffective.

REFERENCES

1. Reaven, G.M., The metabolic syndrome: requiescat in pace. *Clin. Chem.* 2005; 51:931–938.
2. McLaughlin, T. et al., Use of metabolic markers to identify overweight individuals who are insulin resistant. *Ann. Intern. Med.* 2003; 139:802–809.
3. LaMonte, M.J. et al., Cardiorespiratory fitness is inversely associated with the incidence of metabolic syndrome: a prospective study of men and women. *Circulation* 2005; 112:505–512.
4. Mohan, V. et al., Association of physical inactivity with components of metabolic syndrome and coronary artery disease – the Chennai Urban Population Study (CUPS no. 15). *Diabet. Med.* 2005; 22:1206–1211.
5. Koh-Banerjee, P. et al., Prospective study of the association of changes in dietary intake, physical activity, alcohol consumption, and smoking with 9-y gain in waist circumference among 16 587 US men. *Am. J. Clin. Nutr.* 2003; 78:719–727.
6. Dumortier, M. et al,. Low intensity endurance exercise targeted for lipid oxidation improves body composition and insulin sensitivity in patients with the metabolic syndrome. *Diabetes Metab.* 2003; 29:509–518.
7. Mackie, B.D. and Zafari, A.M., Physical activity and the metabolic syndrome. *Hosp. Physician* 2006; 42:26–38.

8. Wallace, M.B., Mills, B.D., and Browing, C.L., Effects of cross-training on markers of insulin resistance/hyperinsulinemia. *Basic Sci. Med. Sci. Sports Exerc.* 1997; 29:1170–1175.

9. Trivedi, M.H. et al., Exercise as an augmentation strategy for treatment of major depression. *J. Psychiatr. Pract.* 2006; 12:205–213.

10. Beebe, L.H. et al., Effects of exercise on mental and physical health parameters of persons with schizophrenia. *Issues Ment. Health Nurs.* 2005; 26:661–676.

11. Mannerkorpi, K., Exercise in fibromyalgia. *Curr. Opin. Rheumatol.* 2005; 17:190–194.

12. Whitelaw, W.A., Brant, R.F., and Flemons, W.W. Clinical usefulness of home oximetry compared with polysomnography for assessment of sleep apnea. *Am. J. Respir. Crit. Care Med.* 2005; 171:188–193.

13. Burns, S.J. et al., A piloty study into the therapeutic effects of music therapy at a cancer help center. *Altern. Ther. Health. Med.* 2001; 7:48–56.

14. Sudsuang, R., Chentanez, V., and Veluvan, K. Effect of Buddhist meditation on serum cortisol and total protein levels, blood pressure, pulse rate, lung volume and reaction time. *Physiol. Behav.* 1991; 50:543–548.

15. Woolery, A. et al., A yoga intervention for young adults with elevated symptoms of depression. *Altern. Ther. Health Med.* 2004; 10:60–63.

16. Jin, P., Changes in heart rate, noradrenaline, cortisol and mood during Tai Chi. *J. Psychosom. Res.* 1989; 33:197–206.

17. Pawlow, L.A. and Jones, G.E., The impact of abbreviated progressive muscle relaxation on salivary cortisol and salivary immunoglobulin A (sIgA). *Appl. Psychophysiol. Biofeedback* 2005; 30:375–387.

18. Berk, L.S. et al., Neuroendocrine and stress hormone changes during mirthful laughter. *Am. J. Med. Sci.* 1989; 298:390–396.

19. Rodondi, N. et al., High risk for hyperlipidemia and the metabolic syndrome after an episode of hypertriglyceridemia during 13-cis retinoic acid therapy for acne: a pharmacogenetic study. *Ann. Intern. Med.* 2002; 136:582–589.

20. Murata, H., Hruz, P.W., and Mueckler, M., The mechanism of insulin resistance caused by HIV protease inhibitor therapy. *J. Biol. Chem.* 2000; 275:20251–20254.

■ INDEX

Printed and bound by CPI Group (UK) Ltd, Croydon, CR0 4YY

08/05/2025

01865017-0001